Ergebnisse der Mathematik und ihrer Grenzgebiete

Band 81

Herausgegeben von P. R. Halmos · P. J. Hilton
R. Remmert · B. Szőkefalvi-Nagy

Unter Mitwirkung von L. V. Ahlfors · R. Baer
F. L. Bauer · A. Dold · J. L. Doob · S. Eilenberg
K. W. Gruenberg · M. Kneser · G. H. Müller
M. M. Postnikov · B. Segre · E. Sperner

Geschäftsführender Herausgeber: P. J. Hilton

J. N. Crossley · Anil Nerode

Combinatorial Functors

Springer-Verlag
Berlin Heidelberg New York 1974

J. N. Crossley, Professor of Mathematics
Monash University, Melbourne, Australia

Anil Nerode, Professor of Mathematics
Cornell University, Ithaca, New York, U.S.A.

AMS Subject Classifications (1970):

02-02, 02 H 15, 02 F 40, 02 F 45, 02 H 99, 18 B 99

ISBN 978-3-642-85935-9   ISBN 978-3-642-85933-5 (eBook)
DOI 10.1007/978-3-642-85933-5

Preface

nullane de tantis gregibus tibi digna videtur?
rara avis in terra nigroque simillima cygno.

Juvenal Sat. VI 161, 165.

1966 – JNC visits AN at Cornell. An idea emerges.
1968 – JNC is at U.C.L.A. for the Logic Year. The Los Angeles manuscript appears.
1970 – AN visits JNC at Monash.
1971 – The Australian manuscript appears.
1972 – JNC visits AN at Cornell. Here is the result.

We gratefully acknowledge support from Cornell University, University of California at Los Angeles, Monash University and National Science Foundation Grants GP 14363, 22719 and 28169. We are deeply indebted to the many people who have helped us. Amongst the mathematicians, we are particularly grateful to J. C. E. Dekker, John Myhill, Erik Ellentuck, Peter Aczel, Chris Ash, Charlotte Chell, Ed Eisenberg, Dave Gillam, Bill Gross, Alan Hamilton, Louise Hay, Georg Kreisel, Phil Lavori, Ray Liggett, Al Manaster, Michael D. Morley, Joe Rosenstein, Graham Sainsbury, Bob Soare and Michael Venning. Last, but by no means least, we thank Anne-Marie Vandenberg, Esther Monroe, Arletta Havlik, Dolores Pendell, and Cathy Stevens and the girls of the Mathematics Department of UCLA in 1968 for hours and hours of excellent typing.

Thanksgiving
November 1972
Ithaca, New York

J. N. Crossley
Anil Nerode

Contents

0. Introduction

Algebra and model theory are concerned with the properties preserved under isomorphisms. Here we study properties preserved under effective isomorphisms. We discuss in turn past, present and future results.

Extant work prior to 1970 is surveyed in Crossley [1970]. For our present motivation see Section 10 below which should be read now (and later). Dekker [1955] introduced the first special case: recursive equivalence types (RETs). The best accounts are Dekker's useful [1966] and Dekker and Myhill's pioneering monograph [1960] (see § 11 below). Next came constructive order types introduced independently by Crossley [1963] and Manaster (unpublished) (see Crossley [1969] and § 12 below). Hassett [1964] treated groups and Dekker [1969] and Hamilton [1970] vector spaces. We encompass all these provided finitely generated systems are finite.

Now recursion theory may be regarded as the study of effective uniformities. These earlier recursion theoretic studies were based on uniformities on sets and structures. We assign a fundamental rôle to universal structures which are entirely effectively presented. We can then identify all other structures of that type with substructures of the universal one. The uniformities inherited by the substructures then smooth the way. But most important we develop uniformities on morphisms. Aczel's work [1966] on continuous homomorphisms and RETs first indicated this view was profitable. The additional strength here yields very general results that were previously unavailable.

In Part I we introduce the appropriate categories and define combinatorial functors. In Part II we treat the elementary algebra and model theory needed to analyze the universal structures. The important model theoretic notion of dimension (Marsh [1966]) brings the class of finitary strict combinatorial functors into focus (Part III) for these can be handled by arithmetical techniques akin to those of Myhill [1958] and Nerode [1961] for RETs. Effective notions are only introduced in Part IV. We find that separation of classical and effective aspects clarifies proofs (though others may disapprove). After this we treat only suitable categories $\mathbb{C} = \mathbb{C}(\mathfrak{M})$ arising from a fixed fully effective $\aleph_0$-categorical model $\mathfrak{M}$. Subsequent theorems bear out our claim that this is a natural generalization of previously studied special cases. The fundamental notion is recursive equivalence of algebraically closed sets by isomorphisms extendable to one-one partial recursive functions. The equivalence classes are the $\mathbb{C}$-RETs. The fundamental tool is the class of partial recursive combinatorial functors F. These induce functions F on $\mathbb{C}$-RETs. We concentrate on Dedekind $\mathbb{C}$-RETs: those whose representatives cannot be mapped onto a proper subset by a one-one partial recursive function. We thus study structures whose elementary relations are all (restrictions of) wholly effective relations but whose domains are (effectively) finite in Dedekind's sense. In Part V our main concern is with equations $\mathsf{F} = \mathsf{G}$ in $\mathbb{C}$-RETs when solutions are plentiful. In Part VI we introduce chains of morphisms and their chain types. (0-chains are isomorphism types.) The theorems show how equations $\mathsf{F} = \mathsf{G}$ yield relations among chain types which in turn give rise to relations on $\mathbb{C}$-RETs. This has the theory of extensions to isols (Nerode [1961]) as its easiest special case.

With a simple dimension present solutions of $\mathsf{F} = \mathsf{G}$ in $\mathbb{C}$-RETs (F, G finitary recursive strict combinatorial functors) are determined by finite 0-chain solutions. This allows us to determine truth conditions for universal sentences (Part VII) built up from such equations. Much of the theory of dimension classically depends on computations with independent sets. Sound types are defined so that such computations succeed. They are the $\mathbb{C}$-RETs with a representative having a basis extendable to a recursively enumerable independent set (Dekker [1969]). We then obtain a quite satisfactory theory of functions on $\mathbb{C}$-RETs to sound Dedekind types (Part VIII) and a tight relation between 0-chain type extensions and strict combinatorial functors. In Part IX we turn to suitable categories with an algebraic property weaker than dimension. Amongst these are categories of sets, vector spaces and linear orderings. Here the regressive $\mathbb{C}$-RET solutions of $\mathsf{F} = \mathsf{G}$ (F, G finitary recursive combinatorial functors) are determined by the finite 1-chain type solutions.

The culmination of the work (Part X) is a simultaneous satisfiability criterion for extensions to $\mathbb{C}$-RETs of relations on chain types. It is the principal justification for the whole extension procedure and has as particular cases many disparate constructions of many authors. An important application is the embeddability of diophantine correct models of formal arithmetic in the RETs.

For the future we observe that the priority method will yield results for very narrow subcategories generalizing those of Hay [1966] for cosimple sets, and that forcing will yield results for choiceless models of set theory generalizing those of Ellentuck [1965, 66, 69]. The development of combinatorial functors of Part I is far more general than subsequent parts require and forms the basis for investigation of the case when finitely generated systems may be infinite. But these developments lie in the future.

Part I. Categories and Functors

1. Categories

This monograph concerns certain categories $\mathbb{C}$ equipped with a full subcategory $°\mathbb{C}$ with "small" or "finitely generated" objects. Both of these categories are entirely concrete. In particular the objects $\mathfrak{A}, \mathfrak{B}, \mathfrak{C}, \ldots$ are sets and the morphisms $p, q, r, \ldots$ are maps.

Suppose we define $U(\mathfrak{A}) = \{\mathfrak{B} : \mathfrak{B}$ is an object of $\mathbb{C}$ and $\mathfrak{B} \supseteq \mathfrak{A}\}$ for each object $\mathfrak{A}$ of $°\mathbb{C}$. Assume that the $U(\mathfrak{A})$ together with the empty set $\emptyset$ are closed under finite intersections. These sets then form a base (the *standard base*) for the *weak topology* on objects.

The definition of combinatorial functor $F: \mathbb{C} \to \mathbb{C}'$ will entail that F preserves inclusions and the inverse image under F of a standard base set is a standard base set.

Thus we study a particular class of continuous, inclusion preserving functors on concrete categories.

We shall use the term map in the following sense. A *map* p is a triple (A, f, B) where A, B are sets and f is a single valued set of ordered pairs, A is the set of first elements of ordered pairs in f and B contains the set of second elements of ordered pairs in f. We write $A = \text{dom } p$, $f = \text{graph } p$ and $B = \text{codom } p$. The range of p, $\text{ran } p$, is the set of second elements of ordered pairs in graph p and as usual we write $\text{ran } p = p(\text{dom } p)$ instead of $\text{ran } p = \text{graph } p(\text{dom } p)$. We shall frequently write $p \subseteq q$ if p, q are both maps and the three set-theoretic inclusions

$$\text{dom } p \subseteq \text{dom } q, \quad \text{graph } p \subseteq \text{graph } q \quad \text{and} \quad \text{codom } p \subseteq \text{codom } q$$

hold. All morphisms will be maps in the above sense.

Now we proceed to precise specification of the categories. The composition $p \circ q$ of morphisms p and q (generally written $p\,q$) is defined if $\text{codom } q = \text{dom } p$. then $\text{dom}(p \circ q) = \text{dom } q$, $\text{codom}(p \circ q) = \text{codom } p$, and $\text{graph}(p \circ q) = \text{graph } p \circ \text{graph } q$. We shall need two operations on maps. Given a set $\{p^i : i \in I\}$ of maps then $\bigcap \{\text{graph } p^i : i \in I\}$ may or may not be a map from $\bigcap \{\text{dom } p^i : i \in I\}$ into $\bigcap \{\text{codom } p^i : i \in I\}$. If it is we say that $\{p^i : i \in I\}$ has an *intersection* and write this map as $\bigcap \{p^i : i \in I\}$.

Clearly an intersection, if defined, is unique. Thus if it is *defined*

$$\bigcap \{p^i: i \in I\} = (\bigcap_{i \in I} \text{dom } p^i, \bigcap_{i \in I} \text{graph } p^i, \bigcap_{i \in I} \text{codom } p^i).$$

We observe that this intersection is simply co-ordinatewise intersection of the appropriate triples of sets. We also note that

$$p \subseteq q \quad \text{if, and only if,} \quad p \cap q = p$$

(where, as usual, we write $p \cap q$ for $\bigcap \{p, q\}$).

The other operation we require is that of direct union. A collection $\mathscr{P} = \{p^i: i \in I\}$ of maps is said to be *directed* (by inclusion) if, given $p^i, p^j \in \mathscr{P}$, there exists $p^k \in \mathscr{P}$ such that $p^i \subseteq p^k$ and $p^j \subseteq p^k$. If $\mathscr{P} = \{p^i: i \in I\}$ is a directed set of maps then the directed union of $\mathscr{P}$,

$$\bigcup \{p^i: i \in I\} = (\bigcup_{i \in I} \text{dom } p^i, \bigcup_{i \in I} \text{graph } p^i, \bigcup_{i \in I} \text{codom } p^i).$$

So again, $\bigcup \{p^i: i \in I\}$ is obtained, if defined, by taking co-ordinatewise unions of triples.

We shall eventually list the requirements for a category to be appropriate for combinatorial functors. But first we give some examples which we shall frequently use in the sequel and which are adequate to give the flavour of the intended idea. As above, the objects are sets and the morphisms are maps with composition as defined above. In fact all morphisms will be one-one maps. As usual we identify objects and the identity maps on them but in no sense do we think of isomorphic objects as indistinguishable.

Example 1.1. The category $\mathbb{S}$ (sets).

The objects are sets of natural numbers and the morphisms one-one maps·between sets of natural numbers. $^\circ\mathbb{S}$ is the full subcategory whose objects are *finite* sets of natural numbers.

This is the category of sets which we shall use later for the (original) recursive equivalence types. In set theory another example is afforded by taking $\mathbb{C}$ as the category whose objects are all sets and whose morphisms are all one-one maps. Here $^\circ\mathbb{C}$ is the full subcategory of finite sets.

Example 1.2. The category $\mathbb{L}$ (linear orderings). The objects are sets of rational numbers, the morphisms are one-one order preserving maps between objects. $^\circ\mathbb{L}$ is the full subcategory of finite sets of rationals.

This is the category which yields the constructive order types.

In set theory in general we could consider a universal linear ordering of the universe. Then take as objects of the category arbitrary subsets and as morphisms one-one order preserving maps.

Example 1.3. The category $\mathbb{B}$ (Boolean algebras). The objects are those subsets of a fixed countable atomless Boolean algebra which are non-empty and closed under the Boolean operations (subalgebras). The morphisms are Boolean monomorphisms between objects. $°\mathbb{B}$ has as objects the sets which are finite (or equivalently, finitely generated) subalgebras.

Example 1.4. The category $\mathbb{V}$ (vector spaces). The objects are those subsets of a fixed countably infinite dimensional vector spaces over a fixed field K which are subspaces. The morphisms are vector space linear transformations between objects. $°\mathbb{V}$ is the full subcategory whose objects are the sets which are finite dimensional subspaces.

Example 1.5. The category $\mathbb{ON}$ (ordinals). The objects are initial segments of the ordinals. The morphisms are the inclusion maps between ordinals. $°\mathbb{ON}$ is the full subcategory of successor ordinals. (Note 'hat in this case it may happen that $p \subseteq q \in °\mathbb{ON}$ and yet $p \notin °\mathbb{ON}$.)

Finally we give a general example (which does not however include many relational systems and in particular does not cover the case of linear orderings).

Example 1.6. The category $\mathbb{A}$ (universal algebra). The objects are subsets closed under the operations of a fixed algebra with finitary operations (subalgebras). The morphisms are monomorphisms between objects. $°\mathbb{A}$ is the full subcategory whose objects are sets which are finitely generated subalgebras.

All the categories above have common features which allow the development of a theory of combinatorial functors. These features are listed as (1.1)–(1.11) below. These will be the basis for the next section. We verify that any universal algebra category $\mathbb{A}$ has these properties but leave the verification for the other categories to the reader.

(1.1) *Concreteness.* The objects of $\mathbb{C}$ are sets. The morphisms of $\mathbb{C}$ are maps between objects of $\mathbb{C}$. Composition in the category is composition of maps.

(1.2) *Injections.* All morphisms of $\mathbb{C}$ are one-one maps.

In $\mathbb{A}$ if a monomorphism p has an inverse p^{-1}, then p^{-1} is also a monomorphism.

(1.3) *Inverses.* If p is a morphism of $\mathbb{C}$ and p, as a map, has inverse p^{-1}, then p^{-1} is a morphism of $\mathbb{C}$.

In $\mathbb{A}$ if $\mathfrak{A}, \mathfrak{B}$ are subalgebras and $\mathfrak{A} \subseteq \mathfrak{B}$ then $\mathfrak{A}$ is in fact a subalgebra of $\mathfrak{B}$ so the inclusion map of $\mathfrak{A}$ in $\mathfrak{B}$ is a monomorphism.

(1.4) *Inclusions.* If $\mathfrak{A}, \mathfrak{B}$ are objects of $\mathbb{C}$ and $\mathfrak{A}$ is a subset of $\mathfrak{B}$ then the inclusion map $\mathfrak{A} \to \mathfrak{B}$ is a morphism of $\mathbb{C}$.

For algebras the intersection of monomorphisms is again a monomorphism and the intersection of subalgebras is a subalgebra.

(1.5) *Intersections.* If a set of morphisms of $\mathbb{C}$ has an intersection (as a set of maps) then that intersection is in $\mathbb{C}$.

Note that in the case that the morphisms are identity morphisms, that is, objects, then the intersection is always defined. So for any category $\mathbb{C}$ whose morphisms are maps in our sense, the intersection of *any* set of objects of $\mathbb{C}$ is in $\mathbb{C}$ if $\mathbb{C}$ satisfies (1.5).

Because every set of *objects* has an intersection it follows that if $\{\mathfrak{A}^i\colon i\in I\}$ is a set of objects and if there exists an object $\mathfrak{A}$ with $\mathfrak{A}^i\subseteq\mathfrak{A}$ for all $i\in I$ then there is a smallest $\mathfrak{A}$ with this property. We write $\bigvee\{\mathfrak{A}^i\colon i\in I\}$ for this $\mathfrak{A}$. Thus

(1) $$\bigvee\{\mathfrak{A}^i\colon i\in I\}=\bigcap\{\mathfrak{B}\colon \mathfrak{A}^i\subseteq\mathfrak{B} \text{ for all } i\in I\}.$$

For morphisms it is not the case that every set has an intersection. (Consider the morphisms in $\mathbb{L}$ given by $p(0)=0$, $p(1)=2$ and $q(0)=1$, $q(2)=2$.) However, if the intersection corresponding to (1) exists we write

$$\bigvee\{p^i\colon i\in I\}=\bigcap\{q\colon p^i\subseteq q \text{ for all } i\in I\}.$$

In the case of the category $\mathbf{A}$, $\bigvee\{\mathfrak{A}^i\colon i\in I\}$ is the subalgebra generated by the set theoretic union of the $\mathfrak{A}^i$.

(1.6) *Directed unions.* The union of any set of morphisms of $\mathbb{C}$, directed under inclusion, is a morphism of $\mathbb{C}$.

Proof for $\mathbf{A}$. Since the operations of $\mathbf{A}$ are finitary consider an operation $\omega(x_0,\ldots,x_{n-1})$. If $a_0,\ldots,a_{n-1}$ are in the union of the domains of morphisms $p^0,\ldots,p^{n-1}$ and the morphisms are directed under inclusion there is (by induction) a morphism p such that $a_i\in\text{dom }p$ for $i<n$, and $p(a_i)=p^{n_i}(a_i)$ for $a_i\in\text{dom }p^{n_i}$. It now follows easily that the union of the monomorphisms is a monomorphism. $\square$

As a particular case of (1.6) we immediately have

(1.6') The objects of $\mathbb{C}$ are closed under formation of directed unions.

Now we turn to properties of $°\mathbb{C}$. In the case of a universal algebra category $\mathbf{A}$ the finitely generated objects in $\mathbf{A}$ are the objects of the full subcategory $°\mathbf{A}$.

(1.7) *Subcategory.* $°\mathbb{C}$ is a full subcategory of $\mathbb{C}$.

Thus if p is a morphism of $°\mathbb{C}$ then dom p and codom p are objects of $°\mathbb{C}$.

The next property is very important.

(1.8) *Restriction.* If p is a morphism of $\mathbb{C}$, $\mathfrak{A}$ is an object of $\mathbb{C}$ with $\mathfrak{A}\subseteq\text{dom }p$, then the *bijection* $q\subseteq p$ with domain $\mathfrak{A}$ is a morphism of $\mathbb{C}$. If $\mathfrak{A}\in°\mathbb{C}$ and $p\in\mathbb{C}$ then $q\in°\mathbb{C}$.

Proof for $\mathbb{C} = \mathbf{A}$ where $^\circ\mathbf{A}$ has as objects the finitely generated sub-algebras. If $\mathfrak{A} \subseteq \mathrm{dom}\, p$ then restricting p to $\mathfrak{A}$ gives a monomorphism. Restricting the codomain of p to the image of $\mathfrak{A}$ under p gives the required bijection. If $\mathfrak{A}$ is finitely generated then the image of $\mathfrak{A}$ under p is finitely generated (by the images under p of the generators of $\mathfrak{A}$) so the bijection q, being a morphism between objects of $^\circ\mathbf{A}$ in a full sub-category, is in $^\circ\mathbf{A}$. $\square$

As a corollary of restriction (1.8) and inclusion (1.4) it follows that any morphism can be factored into an isomorphism and an inclusion map (with both in $\mathbb{C}$ or both in $^\circ\mathbb{C}$ as appropriate).

Any subalgebra is the directed union of its finitely generated sub-algebras (or equivalently the finitely generated subalgebras contained in it). Any monomorphism in $\mathbf{A}$ is therefore a directed union of its restrictions (in domain and codomain) to finitely generated subalgebras.

(1.9) *Approximation.* If p is any morphism of $\mathbb{C}$, then the morphisms p' of $^\circ\mathbb{C}$ with $p' \subseteq p$ form a directed set with union p.

Again, specializing to objects: any object is the directed union of the objects of $^\circ\mathbb{C}$ contained in it.

Approximation allows us to approach infinite objects; the next property allows us to build up objects in $^\circ\mathbb{C}$ to some extent. We say that a finite sequence $p, q, \ldots$ of morphisms is *compatible in* $\mathbb{C}$ if there is a morphism r in $\mathbb{C}$ with $p \subseteq r, q \subseteq r, \ldots$.

(1.10) *Compatibility.* If p, q are morphisms of $^\circ\mathbb{C}$, compatible in $\mathbb{C}$, then there is (under $\subseteq$) a least morphism r in $\mathbb{C}$ such that $p, q \subseteq r$, and moreover, $r \in {}^\circ\mathbb{C}$.

Proof for $\mathbb{C} = \mathbf{A}$. Let A^0 generate $\mathrm{dom}\, p$ and B^0 generate $\mathrm{dom}\, q$ and similarly let A^1, B^1 generate $\mathrm{codom}\, p$ and $\mathrm{codom}\, q$. Then $A^0 \cup B^0$ generates a subalgebra $\mathfrak{C}$. If $p, q \subseteq r$ then $\mathrm{dom}\, r \supseteq \mathfrak{C}$. Let r^0 be r with domain restricted to $\mathfrak{C}$. Now let $A^1 \cup B^1$ generate $\mathfrak{D}$ then $r^0(\mathfrak{C}) \subseteq \mathfrak{D}$. Let $r' = (\mathfrak{C}, \mathrm{graph}\, r^0, \mathfrak{D})$. Clearly r' is the least morphism as required. More-over, $r' \in {}^\circ\mathbf{A}$ since $\mathfrak{C}, \mathfrak{D}$ are objects of $^\circ\mathbf{A}$. $\square$

As an immediate corollary of (1.10) we obtain by induction on n:

(1.10') If $p^0, \ldots, p^n$ are morphisms of $^\circ\mathbb{C}$, compatible in $\mathbb{C}$, then there is a least morphism r in $\mathbb{C}$ such that $p^0, \ldots, p^n \subseteq r$, and moreover, $r \in {}^\circ\mathbb{C}$.

In this case we write $r = \bigvee_{i=0}^{n} p^i$ and note that $\bigvee$ does not, in general, denote union.

The last property we consider allows us to reduce considerations of morphisms in $^\circ\mathbb{C}$ to maps which are quite literally finite.

Let f be a map, not necessarily in $\mathbb{C}$, and let p be a morphism in $\mathbb{C}$. We shall say that f *generates* p if (i) $f \subseteq p$ and (ii) whenever q is a morphism of $\mathbb{C}$ and $f \subseteq q$ then $p \subseteq q$. A morphism p of $\mathbb{C}$ is said to be *finitely generated* if for some finite map f, f generates p. (To say f is finite means its graph and codomain are finite sets.)

(1.11) *Finite generation.* Every morphism in $^{\circ}\mathbb{C}$ is finitely generated.

For $\mathbb{C} = \mathbf{A}$ and $^{\circ}\mathbf{A}$ the full subcategory of finitely generated subalgebras the property is easily verified. Let A finitely generate dom p where $p \in {}^{\circ}\mathbf{A}$ and let B finitely generate codom p. Let $f = (A, (\text{graph } p)|A, B)$ then f is a finite map generating p.

The reader will readily check that (1.1) – (1.11) are satisfied by our other examples. In the next section we shall refer directly to (1.1) – (1.11) but on first reading the example of a universal algebra category $\mathbf{A}$ will be a useful paradigm. We shall call a category which satisfies (1.1) – (1.11) an *appropriate* category.

It is useful at this point to consider product categories for thereby we reduce the study of functors of several variables to functors of one variable.

If $\mathbb{C}_0, \ldots, \mathbb{C}_n$ are categories then the product category $\mathbb{C}_0 \times \cdots \times \mathbb{C}_n$ has as objects $(n+1)$-tuples $(\mathfrak{A}_0, \ldots, \mathfrak{A}_n)$ of objects $\mathfrak{A}_i$ from $\mathbb{C}_i$ and its morphisms are $(n+1)$-tuples $(p_0, \ldots, p_n)$ with p_i from $\mathbb{C}_i$. Composition of morphisms is defined co-ordinatewise so $(p_0, \ldots, p_n)(q_0, \ldots, q_n) = (p_0 q_0, \ldots, p_n q_n)$.

Directed union in $\mathbb{C}_0 \times \cdots \times \mathbb{C}_n$ is, co-ordinatewise, directed union in the several categories and we similarly treat inclusion, intersection, etc. The objects and morphisms of $\mathbb{C}_0 \times \cdots \times \mathbb{C}_n$ only satisfy (1.1) in each co-ordinate if each $\mathbb{C}_i$ satisfies (1.1). However if $\mathbb{C}_0, \ldots, \mathbb{C}_n$ satisfy (1.2) – (1.11) then so does $\mathbb{C}_0 \times \cdots \times \mathbb{C}_n$. With this interpretation of (1.1) for $\mathbb{C}_0 \times \cdots \times \mathbb{C}_n$ it follows that if $\mathbb{C}_0, \ldots, \mathbb{C}_n$ are appropriate then so is $\mathbb{C}_0 \times \cdots \times \mathbb{C}_n$.

Since everything is done co-ordinatewise we shall just write $\mathbb{C}$ for $\mathbb{C}_0 \times \cdots \times \mathbb{C}_n$ and a *subscript will always indicate a co-ordinate* throughout the remainder of the book, unless otherwise indicated.

Thus whenever A is an n-tuple then A_i will denote the i-th co-ordinate whether A be an n-tuple of sets, morphisms, objects or whatever. We shall not in general draw explicit attention to this fact. The general rule is always: Do everything co-ordinatewise.

2. Morphism Combinatorial Functors

We shall consider only categories satisfying (1.1) – (1.11) so in particular the category $\mathbf{A}$ of subalgebras of an algebra will be such. By a functor $F: \mathbb{C} \to \mathbb{C}'$ we mean a function which assigns to each morphism p of $\mathbb{C}$

a morphism Fp of $\mathbb{C}'$ such that $F(p \circ q) = Fp \circ Fq$. We shall, however, always use "functor" to mean a function as above such that

(1) if $\mathfrak{A}$ is an object, $F\mathfrak{A}$ is an object,

(2) if p is an identity morphism then Fp is an identity, and

(3) if p is an inclusion map $p: \mathfrak{A} \subseteq \mathfrak{B}$ then Fp is the inclusion map $Fp: F\mathfrak{A} \subseteq F\mathfrak{B}$.

Definition 2.1. A functor $F: \mathbb{C} \to \mathbb{C}'$ is said to be *morphism combinatorial* if for any morphism q in $^\circ\mathbb{C}'$ there exists p in $^\circ\mathbb{C}$ such that for all r in $\mathbb{C}$, if $q \subseteq Fs$ for some $s \in \mathbb{C}$ then

$(*)$ $\qquad\qquad\qquad q \subseteq Fr$ if, and only if, $p \subseteq r$.

In this case we write $p = F^\leftarrow q$. $F^\leftarrow q$ is the *pseudo-inverse* of q.

We observe $(*)$ can be written

$$q \subseteq Fr \quad \text{if and only if} \quad F^\leftarrow q \subseteq r.$$

The combinatorial property is best described topologically. If $p \in {}^\circ\mathbb{C}$ let $U(p) = \{q \in \mathbb{C}: p \subseteq q\}$. The weak topology on $\mathbb{C}$ is the smallest topology declaring $U(p)$ open for all $p \in {}^\circ\mathbb{C}$. By the compatibility property (1.10′) if $p^0, \ldots, p^n \in {}^\circ\mathbb{C}$ then $U(p^0) \cap \cdots \cap U(p^n) \neq \emptyset$ implies $p^0, \ldots, p_n$ are compatible in $\mathbb{C}$ so there is a smallest q in $\mathbb{C}$ with $p^0, \ldots, p^n \subseteq q$ and $q \in {}^\circ\mathbb{C}$. But then $U(p^0) \cap \cdots \cap U(p^n) = U(q)$. Hence the $U(p)$ for $p \in {}^\circ\mathbb{C}$ together with the empty set form a base for the weak topology. We call this base the *standard* base.

Now if a functor $F: \mathbb{C} \to \mathbb{C}'$ is morphism combinatorial then the inverse image of a standard base set in $\mathbb{C}'$ is a standard base set in $\mathbb{C}$. Thus morphism combinatorial functors are a special class of continuous functors.

Theorem 2.2. *Morphism combinatorial functors are closed under composition.*

Proof. For functors of one variable the theorem is immediate from the topological characterization. Consider now a composition $K(p) = H(F(p), G(p))$ where $F: \mathbb{C}^2 \to \mathbb{C}^0$, $G: \mathbb{C}^2 \to \mathbb{C}^1$ and $H: \mathbb{C}^0 \times \mathbb{C}^1 \to \mathbb{C}^3$. If $q \subseteq H(F(p), G(p))$ then $H^\leftarrow(q) \subseteq (F(p), G(p))$ hence $F^\leftarrow(H^\leftarrow(q))_0 \subseteq p$ and $G^\leftarrow(H^\leftarrow(q))_1 \subseteq p$. From this by the compatibility property there is a least morphism r (in $^\circ\mathbb{C}^2$) such that $F^\leftarrow(H^\leftarrow(q))_0 \subseteq r$ and $G^\leftarrow(H^\leftarrow(q))_1 \subseteq r$. Set $K^\leftarrow(q) = r$. We now leave the reader to check that $K^\leftarrow(q)$ so defined has the required property. The general case is a notational complication of the above argument and we also leave that to the reader. $\qquad \square$

A functor F is said to be *monotone* if $p \subseteq p'$ implies $Fp \subseteq Fp'$. A functor F is said to *preserve direct unions* if whenever $\{p^i: i \in I\}$ is a directed set of morphisms with $p=\bigcup\{p^i: i\in I\}$ then $\{Fp^i: i\in I\}$ is a directed set of morphisms with (direct) union Fp.

Theorem 2.3. *Suppose $F: \mathbb{C}\to\mathbb{C}'$ is a* function *continuous in the weak topology on morphisms. Then F is monotone on morphisms and preserves direct unions of morphisms.*

Proof. By the approximation property (1.9) for $\mathbb{C}'$ in order to establish monotonicity it suffices to show that if p^1, $p^2\in\mathbb{C}$ with $p^1\subseteq p^2$, $q\in{}^\circ\mathbb{C}'$ and $q\subseteq Fp^1$ then $q\subseteq Fp^2$.

Since F is continuous there is a base set $U(p)$ in $\mathbb{C}$, containing p^1, such that F maps $U(p)$ into $U(q)$. Now $p^1\in U(p)$ implies $p\subseteq p^1$ whence $p\subseteq p^2$. Therefore $p^2\in U(p)$, so $F(p^2)\in U(q)$ and finally therefore $q\subseteq F(p^2)$.

Now we show F preserves direct unions. Let $\{p^i: i\in I\}$ be a directed set of morphisms. Then $\{Fp^i: i\in I\}$ is a directed set since F is monotone. If $p=\bigcup\{p^i: i\in I\}$ then $p^i\subseteq p$ so $Fp^i\subseteq Fp$ for all $i\in I$.

This proves $\bigcup\{Fp^i: i\in I\}\subseteq Fp$. We now prove the opposite inclusion. Suppose $q\subseteq Fp$ where $q\in{}^\circ\mathbb{C}'$. Since F is continuous there is a neighbourhood $U(p')$ of p which is mapped into $U(q)$ by F. By the finite generation property there is a finite map f such that p' is the smallest morphism in ${}^\circ\mathbb{C}$ (or indeed $\mathbb{C}$) such that $f\subseteq p'$. Since $f\subseteq p'\subseteq p$ and p is the directed union of $\{p^i: i\in I\}$, f is finite implies there exists $i^q\in I$ such that $f\subseteq p^{i^q}$. But f finitely generates p' so $f\subseteq p^{i^q}$ implies $p'\subseteq p^{i^q}$. Hence $p^{i^q}\in U(p')$. By the choice of p', it follows that $Fp^{i^q}\in U(q)$ and hence $q\subseteq Fp^{i^q}$. Thus if $q\in{}^\circ\mathbb{C}'$ and $q\subseteq Fp$ then $q\subseteq\bigcup\{Fp^i: i\in I\}$. By the approximation property (1.9) for $\mathbb{C}'$ it follows that $Fp\subseteq\bigcup\{Fp^i: i\in I\}$ and this completes the proof. $\square$

Corollary 2.4. *Morphism combinatorial functors are monotone and preserve direct unions (for both morphisms and objects).* $\square$

Theorem 2.5 (Extension of continuous functions). *Let $F: {}^\circ\mathbb{C}\to\mathbb{C}'$ be a function continuous on morphisms. Then there is a unique function $G: \mathbb{C}\to\mathbb{C}'$ which is continuous on morphisms and extends F.*

Proof. To say that F is continuous means it is continuous in the relative topology induced on ${}^\circ\mathbb{C}$ by the weak topology on $\mathbb{C}$. As in the proof of Theorem 2.3 it follows that F is monotone on ${}^\circ\mathbb{C}$.

By the second part of Theorem 2.3, if G is continuous G will preserve directed unions. Now F is monotone so $\{p': p'\subseteq p\ \&\ p'\in{}^\circ\mathbb{C}\}$ being directed implies $\{Fp': p'\subseteq p\ \&\ p'\in{}^\circ\mathbb{C}\}$ is directed. By the directed union property (1.6) we have $\bigcup\{Fp': p'\subseteq p\ \&\ p'\in{}^\circ\mathbb{C}\}$ is in $\mathbb{C}'$ so we set

$$(4)\qquad\qquad Gp=\bigcup\{Fp': p'\subseteq p\ \&\ p'\in{}^\circ\mathbb{C}\}.$$

Since F is monotone, G extends F and clearly G is monotone.

If G is continuous then (4) holds so G is unique. So we now only need to show that G is continuous. Let $U(q)$ be a basic open set containing Gp. By the finite generation property (1.11), there is a finite map $f \subseteq q$ such that q is the smallest morphism of $\mathbb{C}'$ containing f. Note that $Gp \in U(q)$ implies $f \subseteq q \subseteq Gp$, so $f \subseteq Gp$. From the definition of G, $x \in \mathrm{dom}\, f$ implies there exists a morphism $p^{(x)}$ in $°\mathbb{C}$ such that $p^{(x)} \subseteq p$ and $x \in \mathrm{dom}\, Fp^{(x)}$. Similarly for each $y \in \mathrm{codom}\, f$ there exists p^y in $°\mathbb{C}$ such that $p^y \subseteq p$ and $y \in \mathrm{codom}\, Fp^y$. In the same way for each ordered pair (a, b) in $\mathrm{graph}\, f$ there exists $p(a, b)$ in $°\mathbb{C}$ such that $p(a, b) \subseteq p$ and $(a, b) \in \mathrm{graph}\, Fp(a, b)$. Since f is finite the collection of all such $p^{(x)}$, p^y and $p(a, b)$ is a finite set of morphisms of $°\mathbb{C}$ all contained in p and therefore compatible. By the compatibility property (1.10) for $\mathbb{C}$ there is a smallest morphism p^0 in $\mathbb{C}$ containing all $p^{(x)}$, p^y and $p(a, b)$ and $p^0 \in °\mathbb{C}$. We therefore have $p^0 \subseteq p$ so $p \in U(p^0)$. Finally we prove that G maps $U(p^0)$ into $U(q)$ which will complete the proof that G is continuous. Now G is monotone, so it suffices to show $q \subseteq Gp^0$.

Since G extends F and is monotone, $Gp^{(x)} \subseteq Gp^0$, $Gp^y \subseteq Gp^0$ and $Gp(a, b) \subseteq Gp^0$ for $x \in \mathrm{dom}\, f$, $y \in \mathrm{codom}\, f$ and $(a, b) \in \mathrm{graph}\, f$. But then $f \subseteq Gp^0$. Since q is the smallest morphism containing f, it follows that $q \subseteq Gp^0$. This completes the proof. $\quad\square$

Lemma 2.6. *Suppose* $p, q \in \mathbb{C}$ *and* $p \circ q$ *is defined. Then for all* $h \in °\mathbb{C}$, $h \subseteq p \circ q$ *if, and only if, there exist* $p', q' \in \mathbb{C}$ *such that* $p' \circ q'$ *is defined,* $p' \subseteq p$, $q' \subseteq q$ *and* $h = p' \circ q'$.

(Note that the p', q' of the lemma are not unique.)

Proof. The implication from right to left is trivial. For the converse suppose $h \in °\mathbb{C}$ and $h \subseteq p \circ q$. Then $\mathrm{dom}\, h \subseteq \mathrm{dom}\, q$. By the restriction property (1.8) there is an isomorphism $q' \subseteq q$ with $\mathrm{dom}\, q' = \mathrm{dom}\, h$ and $q' \in °\mathbb{C}$. Then $q' \in °\mathbb{C}$ implies $\mathrm{codom}\, q' \in °\mathbb{C}$. Again by the restriction property the isomorphism p'' with $\mathrm{dom}\, p'' = \mathrm{codom}\, q'$ and $p'' \subseteq p$ is in $°\mathbb{C}$. But then $\mathrm{codom}\, p'' \in °\mathbb{C}$. Also, since $h \subseteq p \circ q$, $\mathrm{codom}\, p'' \subseteq \mathrm{codom}\, h$ by construction. Since $\mathrm{codom}\, p''$ and $\mathrm{codom}\, h$ are in $°\mathbb{C}$ the inclusion map $i: \mathrm{codom}\, p'' \to \mathrm{codom}\, h$ is in $\mathbb{C}$ and hence in $°\mathbb{C}$, since $°\mathbb{C}$ is a full subcategory of $\mathbb{C}$ (by (1.7)). But now i, $p'' \in °\mathbb{C}$ so $i \circ p'' \in °\mathbb{C}$. Set $p' = i \circ p''$ then $h = p' \circ q'$ and by construction $p' \subseteq p$ and $q' \subseteq q$. $\quad\square$

Theorem 2.7. *Suppose* $F: \mathbb{C} \to \mathbb{C}'$ *is a continuous function and* $F^0 = F$ *restricted to* $°\mathbb{C}$ *is a functor from* $°\mathbb{C}$ *to* $\mathbb{C}'$. *Then* F *is a functor.*

Proof. Since F^0 is a functor, $F^0(\mathrm{dom}\, p') = \mathrm{dom}\, F^0 p'$, and $F^0(\mathrm{codom}\, p') = \mathrm{codom}\, F^0 p'$ for all $p' \in °\mathbb{C}$. We show first that these properties extend to F.

Let $p \in \mathbb{C}$ then p is the directed union of $\{p': p' \in °\mathbb{C} \,\&\, p' \subseteq p\}$. Then $\mathrm{dom}\, p = \bigcup \{\mathrm{dom}\, p': p' \in °\mathbb{C} \,\&\, p' \subseteq p\}$. Since F is continuous, F preserves direct unions by Theorem 2.3, so

$$F(\text{dom } p) = \bigcup \{F(\text{dom } p'): p' \in {}^{\circ}\mathbb{C} \,\&\, p' \subseteq p\}$$
$$= \bigcup \{F^0(\text{dom } p'): p' \in {}^{\circ}\mathbb{C} \,\&\, p' \subseteq p\}$$
$$= \bigcup \{\text{dom } F^0 p': p' \in {}^{\circ}\mathbb{C} \,\&\, p' \subseteq p\}$$
$$\subseteq \text{dom } F p$$

since F is monotone.

Conversely if $q \subseteq \text{dom } F p$ and $q \in {}^{\circ}\mathbb{C}'$ then by the finite generation property (1.11) there is a finite map f which generates q. (f may be taken to be an identity map as q is an identity.) Further we may identify f and dom f. As in the proof of Theorem 2.3 we can find a finite set $\{p^k: k \in K\}$ of elements of ${}^{\circ}\mathbb{C}$ with each $p^k \subseteq p$ such that

$$\text{dom } f \subseteq \bigcup \{\text{dom } F p^k: k \in K\}.$$

Since $\{p': p' \in {}^{\circ}\mathbb{C} \,\&\, p' \subseteq p\}$ is directed and F is monotone there exists $p^0 \in {}^{\circ}\mathbb{C}$ with all $p^k \subseteq p^0, p^0 \subseteq p$ and dom $f \subseteq \text{dom } F p^0$. But then $q \subseteq \text{dom } F p^0$ and since $p^0 \in {}^{\circ}\mathbb{C}$, $q \subseteq F(\text{dom } p^0)$. Finally since F is monotone, $F(\text{dom } p^0) \subseteq F(\text{dom } p)$ so dom $F p \subseteq F(\text{dom } p)$. Thus we do have dom $F p = F \text{dom } p$.

In exactly similar fashion we have codom $F p = F(\text{codom } p)$.

That F preserves inclusion maps is easier. Let p be an inclusion map in $\mathbb{C}$ then $F p = \bigcup \{F p': p' \in {}^{\circ}\mathbb{C} \,\&\, p' \subseteq p\}$ as in Theorem 2.3. But each $F p'$ for $p' \in {}^{\circ}\mathbb{C}$ is an inclusion map since $p' \subseteq p$ is an inclusion map, because p is. The union of inclusion maps is an inclusion map so $F p$ is an inclusion as required.

Now we show that $F(p \circ q) = F p \circ F q$ when $p \circ q$ is defined. Suppose $p \circ q$ is defined then codom $q = \text{dom } p$. But by the first part of the proof F commutes with dom and codom so $F p \circ F q$ is defined. Again by the first part dom $F(p \circ q) = F(\text{dom } p \circ q) = F(\text{dom } q) = \text{dom } F q = \text{dom}(F p \circ F q)$ and similarly codom $F(p \circ q) = \text{codom}(F p \circ F q)$. To complete the proof it therefore suffices to show $F(p \circ q) \subseteq F p \circ F q$. So suppose $r \subseteq F(p \circ q)$ where $r \in {}^{\circ}\mathbb{C}'$. Since F is continuous there exists $h \in {}^{\circ}\mathbb{C}$ with $r \subseteq F(h)$ and $h \subseteq p \circ q$. By Lemma 2.6 there exist $p' \subseteq p$, $q' \subseteq q$ with $p', q' \in {}^{\circ}\mathbb{C}$ and $h = p' \circ q'$. Since F restricted to ${}^{\circ}\mathbb{C}$ is a functor, $F(p' \circ q') = F p' \circ F q'$ so $F h = F p' \circ F q'$. Since F is monotone $F p' \subseteq F p$ and $F q' \subseteq F q$. Hence $r \subseteq F(h)$ implies $r \subseteq F p \circ F q$. We have shown $r \subseteq F(p \circ q)$ implies $r \subseteq F p \circ F q$ for $r \in {}^{\circ}\mathbb{C}'$ so by the approximation property (1.9) it follows that $F(p \circ q) \subseteq F p \circ F q$ and the proof is complete. $\square$

We conclude this phase with a condition frequently used to produce combinatorial functors from functors on ${}^{\circ}\mathbb{C}$.

Corollary 2.8. Let $F: {}^{\circ}\mathbb{C} \to \mathbb{C}'$ be a functor satisfying (∗). Then the unique continuous extension G of F to a function from $\mathbb{C}$ to $\mathbb{C}'$ is a morphism combinatorial functor.

Proof. F takes inclusion maps to inclusion maps. From this and the fact that F is functor it can be seen that for p, q morphisms of $°\mathbb{C}$, $p \subseteq q$ implies $Fp \subseteq Fq$. But this expresses the continuity of F. So Theorems 2.5 and 2.7 yield the corollary. □

Lemma 2.9. *Suppose* $F: \mathbb{C} \to \mathbb{C}'$ *is a morphism combinatorial functor. Then* F *preserves intersections of morphisms.*

Proof. We have previously noted that the intersection of any set of objects (identity maps) exists. So if $\{p^i: i \in I\}$ is a set of identity maps and $q \in °\mathbb{C}$ then $q \subseteq F(\bigcap \{p^i: i \in I\})$ if and only if $F^\leftarrow q \subseteq \bigcap \{p^i: i \in I\}$ if and only if $F^\leftarrow q \subseteq p^i$ for all $i \in I$ if and only if $q \subseteq Fp^i$ for all $i \in I$ if and only if $q \subseteq \bigcap \{Fp^i: i \in I\}$. By the approximation property (1.9) for $\mathbb{C}'$ we therefore have $F \bigcap \{p^i: i \in I\} = \bigcap \{Fp^i: i \in I\}$ (when the p^i are all identities).

It immediately follows that if an intersection $p = \bigcap \{p^i: i \in I\}$ of a set of maps exists (in $\mathbb{C}$) then $\text{dom } Fp = F \text{ dom } p$ and $\text{codom } Fp = F \text{ codom } p$. So to show $\bigcap Fp^i = F \bigcap p^i$ it suffices to prove $\text{graph } Fp = \bigcap \{\text{graph } Fp^i: i \in I\}$. Suppose $q \in °\mathbb{C}'$ and $q \subseteq Fp$ then $F^\leftarrow q \subseteq p$ so $F^\leftarrow q \subseteq p^i$ for all $i \in I$. But then $q \subseteq Fp^i$ for all $i \in I$. By the approximation property for $\mathbb{C}'$, $Fp \subseteq Fp^i$ for all $i \in I$. Hence $\text{graph } Fp \subseteq \text{graph } Fp^i$ and the proof is complete. □

We note that in particular if $\{\mathfrak{A}^i: i \in I\}$ is a set of objects in $\mathbb{C}$ then $F \bigcap \{\mathfrak{A}^i: i \in I\} = \bigcap \{F \mathfrak{A}^i: i \in I\}$.

3. Combinatorial Functors

For the most part we shall be solely concerned with combinatorial functors.

Definition 3.1. A functor $F: \mathbb{C} \to \mathbb{C}'$ continuous on morphisms is said to be *combinatorial* if for any identity $q \in °\mathbb{C}'$ there is an identity $p \in °\mathbb{C}$ such that for all identities $r \in \mathbb{C}$, if $q \subseteq Fs$ for some $s \in \mathbb{C}$ then

$$\text{(*)} \qquad\qquad q \subseteq Fr \quad \text{if and only if} \quad p \subseteq r.$$

In this case we write $p = F^\leftarrow q$.

Let $\text{Ob}(\mathbb{C})$ denote the objects of $\mathbb{C}$ then the standard base for the weak topology on $\mathbb{C}$ yields a standard base for the induced topology on $\text{Ob}(\mathbb{C})$. If F is a functor from $\mathbb{C}$ to $\mathbb{C}'$ let $^{\text{Ob}}F$ denote the restriction of F to the objects of $\mathbb{C}$. A functor $F: \mathbb{C} \to \mathbb{C}'$ continuous on morphisms is therefore combinatorial if and only if $^{\text{Ob}}F: \text{Ob}(\mathbb{C}) \to \text{Ob}(\mathbb{C}')$ has the property that the inverse image of a standard base set is a standard base set.

Example 3.1. Let A be a set and let f be a one-one map of A onto a basis for a vector space U over a field K. Let $\mathbb{S}$ be the category of one-one maps between subsets of A and let $°\mathbb{S}$ be the finite maps of $\mathbb{S}$.

Let $\mathbb{V}$ be the category of monomorphisms between subspaces of U and let $°\mathbb{V}$ be the monomorphisms between finite-dimensional subspaces of U.

Let $V: \mathbb{S} \to \mathbb{V}$ assign to each object $\mathfrak{A}$ of $\mathbb{S}$ the object $V(\mathfrak{A})$ generated by $\{f(a): a \in \mathfrak{A}\}$ and to each morphism p of $\mathbb{S}$ the morphisms $V(p)$ from $V(\text{dom } p)$ to $V(\text{codom } p)$ such that for $a \in \text{dom } p$, $V(p)(f(a))=f(p(a))$. Since $\{f(a): a \in \text{dom } p\}$ is a basis for dom $V(p)$, $V(p)$ is well-defined.

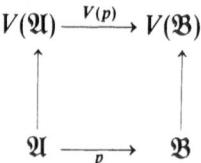

V is obviously a functor. It is combinatorial, but not morphism combinatorial as we now show. Let v be a vector in the space U which is generated by $f(A)$. Then v is a unique linear combination with non-zero coefficients of a finite set $f(A')$ of basis elements, where $A' \subseteq A$. Call A' the *support* of v. Suppose $\mathfrak{A}$ is a finite dimensional subspace of U generated by $v_1, \ldots, v_n$ then $V^{\leftarrow}\mathfrak{A} = \bigcup_{i=1}^{n} \text{support } (v_i)$ satisfies property $(*)$ of Definition 3.1, so V is combinatorial. But V is not morphism combinatorial unless A has only one element (that is, U is one dimensional). For suppose a^1, $a^2 \in A$ with $a^1 \neq a^2$ and let $f(a^1)=b^1$, $f(a^2)=b^2$.

Let $q \in °\mathbb{V}$ be the identity morphism on the one-dimensional subspace of U spanned by b^1+b^2. (Note that q is not in the range of V.) Let p^1: $\{a^1, a^2\} \to \{a^1, a^2\}$ be the identity map then $p^1 \in °\mathbb{S}$. Let p^2: $\{a^1, a^2\} \to \{a^1, a^2\}$ be defined by $p^2(a^1)=a^2$, $p^2(a^2)=a^1$ then $p^2 \in °\mathbb{S}$ also. Now $q \subseteq V p^1$ and $q \subseteq V p^2$ so if $V^{\leftarrow}(q)$ is defined, $V^{\leftarrow}(q) \subseteq p^1 \cap p^2 = \emptyset$, but $q \not\subseteq V(\emptyset)$. So no $V^{\leftarrow}(q)$ can be defined and V is not morphism combinatorial.

We say that F is *continuous* on objects if ^{Ob}F is continuous in the induced topology. So if F is combinatorial, F is continuous on objects.

If all the proofs of Section 2 are modified by restricting to identity maps (objects) the corresponding results are valid for functions F continuous on objects. Since these theorems are the ones actually used in the sequel we state them explicitly (giving them numbers corresponding to those in Section 2).

Theorem 3.2. *Combinatorial functors are closed under composition.*   □

Theorem 3.3. *Suppose* $F: \mathbb{C} \to \mathbb{C}'$ *is a function which is continuous on objects, then* F *is monotone on and preserves direct unions of objects.*   □

Theorem 3.5. *Let* $F: \mathrm{Ob}(°\mathbb{C}) \to \mathrm{Ob}(\mathbb{C}')$ *be continuous. Then there is a unique function* $G: \mathrm{Ob}(\mathbb{C}) \to \mathrm{Ob}(\mathbb{C}')$ *which is continuous on objects and extends* F.   □

The next corollary gives us the existence of extensions of functors from $°\mathbb{C}$ to $\mathbb{C}$.

Corollary 3.8. *Let* $F: °\mathbb{C} \to \mathbb{C}'$ *be a functor satisfying* $\binom{*}{*}$. *Then the unique continuous extension* G *of* F *to a function* $G: \mathbb{C} \to \mathbb{C}'$ *is a combinatorial functor.*

Proof. Similar to that of Corollary 2.8, using still Theorems 2.5 and 2.7. The crucial point is again that any functor $F: °\mathbb{C} \to \mathbb{C}'$ is continuous on morphisms.   □

Lemma 3.9. *Suppose* $F: \mathbb{C} \to \mathbb{C}'$ *is a functor satisfying* $\binom{*}{*}$ *then* F *preserves intersection of objects.*   □

The categories $\mathbb{S}$, $\mathbb{L}$, $\mathbb{V}$, $\mathbb{B}$ and $\mathbb{ON}$ (but not always $\mathbb{A}$) and those later in the book have the property that $°\mathbb{C}$ is closed under arbitrary intersection of objects. We conclude this section with some results particular to this case and much used later. These results apply of course to combinatorial functors.

Theorem 3.10. *Let* $F: \mathbb{C} \to \mathbb{C}'$ *be a functor continuous on objects. Suppose* $°\mathbb{C}$ *is closed under arbitrary intersections of objects. Then* F *satisfies* $\binom{*}{*}$ *if and only if* F *preserves arbitrary intersections of objects.*

Proof. F satisfies $\binom{*}{*}$ implies F preserves intersections of objects in Lemma 3.9 above. For the converse we adopt the convention in this proof that $p, q, r, \ldots$ denote identity maps (objects) only. So suppose F is continuous on objects and preserves arbitrary intersection of objects. If, for a given $q \in °\mathbb{C}'$, there is no $r \in \mathbb{C}$ with $q \subseteq Fr$ then the condition $\binom{*}{*}$ is vacuous. Otherwise by continuity there exists $r' \subseteq r$ with $r' \in °\mathbb{C}$ such that $q \subseteq Fr'$. Let $p = \bigcap \{r': q \subseteq Fr'\}$. By hypothesis $p \in °\mathbb{C}$ and by continuity, if $r \in \mathbb{C}$ and $p \subseteq r$ then $q \subseteq Fp \subseteq Fr$. Conversely, suppose $r \in \mathbb{C}$ and $q \subseteq Fr$, then, by continuity, there exists $r' \in °\mathbb{C}$, such that $r' \subseteq r$ and $q \subseteq Fr'$, but then $p \subseteq r$. So $\binom{*}{*}$ is satisfied.   □

We restate the above criterion in terms of elements.

Theorem 3.11. *Suppose* $F: \mathbb{C} \to \mathbb{C}'$ *is a functor and* $°\mathbb{C}$ *is closed under arbitrary intersections. Then* F *satisfies* $\binom{*}{*}$ *if and only if for all elements* x

we can find an object $F^\leftarrow x \in {}^\circ\mathbb{C}$ such that for all objects $\mathfrak{A}$ of $\mathbb{C}$ and all x

$$x \in F(\mathfrak{A}) \quad \text{if, and only if,} \quad F^\leftarrow x \subseteq \mathfrak{A}.$$

Proof. Suppose F satisfies (${}^*_*$). We need only consider elements x such that $x \in F(\mathfrak{A})$ for some object $\mathfrak{A} \in \mathbb{C}$. By continuity and approximation (1.9) $x \in F(\mathfrak{A}^\circ)$ for some $\mathfrak{A}^\circ \in {}^\circ\mathbb{C}$. Let $F^\leftarrow(x) = \bigcap \{\mathfrak{A}^\circ \in {}^\circ\mathbb{C}: x \in F(\mathfrak{A}^\circ)\}$. Since F is continuous on objects $F^\leftarrow(x)$ has the required property.

Conversely, suppose $F^\leftarrow x$ is given for all elements x and has the required property. Then for all families $\{\mathfrak{A}^i: i \in I\}$ of objects, we have $x \in F(\bigcap \{\mathfrak{A}^i: i \in I\})$ if, and only if, $F^\leftarrow(x) \subseteq \bigcap \{\mathfrak{A}^i: i \in I\}$ if, and only if, $F^\leftarrow(x) \subseteq \mathfrak{A}^i$ for all $i \in I$ if, and only if, $x \in F(\mathfrak{A}^i)$ for all $i \in I$ if, and only if, $x \in \bigcap \{F(\mathfrak{A}^i): i \in I\}$.

Hence $F \cap \{\mathfrak{A}^i: i \in I\} = \bigcap \{F\mathfrak{A}^i: i \in I\}$. We observe that if F preserves intersections then $\mathfrak{A} \subseteq \mathfrak{B}$ implies $F\mathfrak{A} \subseteq F\mathfrak{B}$ since $\mathfrak{A} \cap \mathfrak{B} = \mathfrak{A}$.

Now if we show that F is continuous on objects the result will follow by Theorem 3.11. As in the previous proof $p, q, r, \ldots$ denote identity maps (objects) throughout.

Suppose $q \in {}^\circ\mathbb{C}'$ and $q \subseteq Fr$ where $r \in \mathbb{C}$. By the finite generation property (1.11) there is a finite map f generating q which we may take to be an identity (object). If $x \in f$ then $x \in q \subseteq Fr$ so $x \in Fr$ and then by hypothesis $F^\leftarrow x \subseteq r$ and $F^\leftarrow x \in {}^\circ\mathbb{C}$. Therefore $\{F^\leftarrow x: x \in f\}$ is a finite compatible family contained in an element r of ${}^\circ\mathbb{C}$. By the compatibility property (1.10) for $\mathbb{C}$ there is a smallest object p in $\mathbb{C}$ with all $F^\leftarrow x \subseteq p$ and $p \subseteq r$ and moreover, $p \in {}^\circ\mathbb{C}$. Then by hypothesis $F^\leftarrow x \subseteq p$ implies $x \in Fp$ for all $x \in f$, so $f \subseteq Fp$ and by the definition of finite generation $q \subseteq Fp$. Since F preserves intersections and is therefore monotone; if $p \subseteq r$, $Fp \subseteq Fr$ so $q \subseteq Fr$. Hence $U(p)$ is mapped into the neighbourhood $U(q)$ (in the topology on objects only). Hence F is continuous on objects. Thus the proof is complete. $\square$

We note that $F^\leftarrow$ has been defined now for both elements and morphisms. But if q is an identity morphism then $F^\leftarrow x \subseteq F^\leftarrow q$ for all $x \in q$ and no confusion results.

Theorem 3.12. *Suppose $F: \mathbb{C} \to \mathbb{C}'$ is a combinatorial functor. If p, q are morphisms of $\mathbb{C}$ and $p \cap q$ is defined, then $F(p) \cap F(q)$ is also defined and $F(p) \cap F(q) = F(p \cap q)$.*

Proof. Examine first the case $p \subseteq q$. If $i: \operatorname{dom} p \to \operatorname{dom} q$ and $j: \operatorname{codom} p \to \operatorname{codom} q$ are inclusion maps, then $q \circ i = j \circ p$ and F a functor imply $(Fp) \circ (Fi) = (Fj) \circ (Fp)$. Since Fi, Fj are inclusion maps, we conclude $Fp \subseteq Fq$ as desired. For the general case suppose p, q are morphisms of $\mathbb{C}$ and $p \cap q$ is defined. Then $p \cap q \subseteq p$, $p \cap q \subseteq q$ imply $F(p \cap q) \subseteq Fp$, $F(p \cap q) \subseteq Fq$. We need to know that

$$(\text{codom } F(p)) \cap (\text{codom } F(q)) \subseteq \text{codom } F(p \cap q)$$

and

$$(\text{dom } F(p)) \cap (\text{dom } F(q)) \subseteq \text{dom } F(p \cap q).$$

We confine ourselves to the latter. Suppose $x \in \text{dom } F(p) \cap \text{dom } F(q)$. Since F is a functor, $\text{dom } F(p) = F(\text{dom } p)$, $\text{dom } F(q) = F(\text{dom } q)$, so $x \in F(\text{dom } p) \cap F(\text{dom } q)$. Since F is combinatorial, $F^{\leftarrow} x \subseteq \text{dom } p \cap \text{dom } q = \text{dom } (p \cap q)$. So $x \in F(\text{dom } p \cap q) = \text{dom } F(p \cap q)$. $\square$

Finally we note that by using the same methods as above we can show that, if a functor F satisfies $\overset{*}{(*)}$ and is continuous on objects, then F is continuous on morphisms and therefore a combinatorial functor.

Part II. Model Theory

4. Countable Atomic Models

In the sequel we shall be concerned with categories which arise from countable models of certain first order theories. These include sets, linear orderings, vector spaces over a finite field and some other familiar algebraic structures but we are able to treat these all uniformly. The model theory we present here will be transparent to those familiar for example with Sacks' delightful book [1972] and such readers should read only the theorems and not the proofs in this and the following section. Otherwise the first few chapters of Bell and Slomson [1969] will be sufficient background.

Our main concern here will be with the notion of *algebraic closure* in a model-theoretic sense. The definition we take is the simplest to handle and gives the results we need. It is weaker than some other definitions in the literature but the techniques involved are the same (see Marsh [1966], Morley [1965], Shelah [1971], Jónsson [1962], Baldwin and Lachlan [1971]).

Let L be a first order language with equality. Let T be a complete theory in L. We may assume every formula is equivalent to an atomic formula. For if this is not the case simply use Morley's device [1965] of adding, for each formula $\phi(v_0, \ldots, v_{k-1})$, a new relation symbol R_ϕ and an axiom

$$\forall v_0, \ldots, \forall v_{k-1}[R_\phi(v_0, \ldots, v_{k-1}) \leftrightarrow \phi(v_0, \ldots, v_{k-1})].$$

Then consistency is preserved and every formula in the extended language is provably equivalent (in the extended theory) to a formula of T. We shall assume that T has only infinite models and whenever we treat models (of any theory) we shall assume they are countable (that is, finite or countably infinite).

If $\mathfrak{M} = \langle M, R^i, \ldots \rangle$ is a model for a theory T in the language L then $\mathfrak{M}$ is said to be an *L-structure*. If $\mathfrak{M}$ is an L-structure and $A = \{a_0, a_1, a_2, \ldots\} \subseteq M$ we let $L(A)$ be the language obtained by adding new constants (which we also denote by) $a_0, a_1, a_2, \ldots$ to L. Then the $L(A)$-structure $\langle \mathfrak{M}, A \rangle = \langle \mathfrak{M}, a_0, a_1, \ldots \rangle$ is said to be an *inessential expansion* of $\mathfrak{M}$. If T is the (complete) theory of $\mathfrak{M}$ in the language L then $T(A)$ is defined to be the complete theory of $\langle \mathfrak{M}, A \rangle$ in the language $L(A)$.

We shall write $|\mathfrak{M}|$ for the universe M of the L-structure $\mathfrak{M}$ but we shall often write $A \subseteq \mathfrak{M}$ instead of $A \subseteq |\mathfrak{M}|$.

For each n we let $F_n(T)$ be the set of all formulae with at most $v_0, \ldots, v_{n-1}$ as free variables. We define an equivalence relation $\sim$ on $F_n(T)$ by setting $\phi \sim \phi'$ it $T \vdash \forall v_0 \ldots \forall v_{n-1} (\phi \leftrightarrow \phi')$. The equivalence classes then form a Boolean algebra $B_n(T)$. We shall identity a formula with its equivalence class unless there is ambiguity. Then a formula $\phi \in F_n(T)$ is an *atom* if (the equivalence class of) ϕ is an atom of the Boolean algebra $B_n(T)$.

As usual, two models $\mathfrak{M}^1$, $\mathfrak{M}^2$ are said to be *elementarily equivalent* (written $\mathfrak{M}^1 \equiv \mathfrak{M}^2$) if $\mathfrak{M}^1$, $\mathfrak{M}^2$ have the same theory.

Definition 4.1. A model $\mathfrak{M}$ of T is said to be *atomic* if for each $a_0, a_1, \ldots, a_{n-1} \in \mathfrak{M}$ there is an atom ϕ in $B_n(T)$ such that

$$\mathfrak{M} \vDash \phi [a_0, \ldots, a_{n-1}].$$

Clearly the condition on ϕ in the definition is equivalent to $\phi(a_0, \ldots, a_{n-1}) \in T(\{a_0, \ldots, a_{n-1}\})$. Moreover, if $\mathfrak{M}$ is atomic there is a unique atom ϕ in $B_n(T)$ which $a_0, \ldots, a_{n-1}$ satisfies.

The proof of the second half of the next theorem is by Cantor's "back and forth" argument whose paradigm is the proof that any two countable dense unbordered linear orderings are isomorphic. Since we shall use the argument frequently we include the full details in this case.

Theorem 4.2 (Vaught [1959] Theorems 3.5 and 3.2). *A (complete) theory T has an atomic model if, and only if, each $B_n(T)$ is atomistic. Any two countable atomic models of T are isomorphic.*

Proof. For the first part see Vaught [1959], p. 311. Let $\mathfrak{A}, \mathfrak{B}$ be countable atomic models of T. We shall show that $\mathfrak{A}, \mathfrak{B}$ have the following homogeneity property. Let $a_0, \ldots, a_{n-1} \in \mathfrak{A}$ and $b_0, \ldots, b_{n-1} \in \mathfrak{B}$ and suppose

$$(\mathfrak{A}, a_0, \ldots, a_{n-1}) \equiv (\mathfrak{B}, b_0, \ldots, b_{n-1})$$

then for each $a_n \in \mathfrak{A}$ there exists $b_n \in \mathfrak{B}$ such that

$$(\mathfrak{A}, a_0, \ldots, a_{n-1}, a_n) \equiv (\mathfrak{B}, b_0, \ldots, b_{n-1}, b_n).$$

Now let $a_0, a_2, a_4, \ldots$ and $b_1, b_3, b_5, \ldots$ be enumerations of $\mathfrak{A}, \mathfrak{B}$ (possibly with repetitions). Then by the symmetry (between $\mathfrak{A}$ and $\mathfrak{B}$) of the homogeneity property we can alternately find $b_0, a_1, b_2, a_3, \ldots$ such that for every n

$$(\mathfrak{A}, a_0, a_1, \ldots, a_{n-1}) \equiv (\mathfrak{B}, b_0, b_1, \ldots, b_{n-1}).$$

But then by considering atomic formulae it follows at once that $a_i \to b_i$ is an isomorphism of $\mathfrak{A}$ onto $\mathfrak{B}$.

Now we establish the homogeneity property. Let ϕ be the atom of $B_n(T)$ satisfied by $a_0, \ldots, a_{n-1}$ and let ψ be the atom of $B_{n+1}(T)$ satisfied by $a_0, \ldots, a_{n-1}, a_n$. Since $(\mathfrak{A}, a_0, \ldots, a_{n-1}) \equiv (\mathfrak{B}, b_0, \ldots, b_{n-1})$, ϕ is also satisfied by $b_0, \ldots, b_{n-1}$.

Now ϕ is consistent (with respect to T) with $\exists v_n \psi$ so ϕ is an atom implies $T \vdash \phi \to \exists v_n \psi$. Since $\mathfrak{B} \vDash \phi[b_0, \ldots, b_{n-1}]$ there exists $b_n \in \mathfrak{B}$ such that $\mathfrak{B} \vDash \psi[b_0, \ldots, b_{n-1}, b_n]$, that is $(\mathfrak{B}, b_0, \ldots, b_n) \vDash \psi(b_0, \ldots, b_n)$. Since every sentence in the theory of $(\mathfrak{B}, b_0, \ldots, b_n)$ is a consequence of $\psi(b_0, \ldots, b_n)$, because ψ is an atom, it follows that $(\mathfrak{A}, a_0, \ldots, a_n) \equiv (\mathfrak{B}_0, b_0, \ldots, b_n)$. This completes the proof. $\square$

Now we consider how many possible choices of b_n there were in the proof. There are two cases.

Case (i). There is a natural number k such that

$$(1) \qquad T \vdash \phi(v_0, \ldots, v_{n-1}) \to \exists !^k v_n \psi(v_0, \ldots, v_{n-1})$$

where $\exists !^k v_n$ means "there are exactly k v_n's such that": a phrase clearly expressible in any first order theory with equality. In this case there are at most k choices for b_n.

Case (ii). If Case (i) does not apply. Then there is no k satisfying (1). Hence, for every k,

$$T \vdash \phi(v_0, \ldots, v_{n-1}) \to (\exists^{<k} v_n \psi(v_0, \ldots, v_{n-1}, v_n)$$
$$\vee \exists^{>k} v_n \psi(v_0, \ldots, v_{n-1}, v_n)),$$

where $\exists^{<k}$ means "there are less than k" and $\exists^{>k}$ means "there are more than k". Now T is complete so

$$T \vdash \phi(v_0, \ldots, v_{n-1}) \to \exists^{>k} v_n \psi(v_0, \ldots, v_{n-1}, v_n)$$

for all k. So there are an infinite number of possible b_n.

The b_n occurring under Case (i) will be said to be "algebraic" over $b_0, \ldots, b_{n-1}$. Formally we proceed as follows.

A subset X of $\mathfrak{M}$ is said to be *definable* from $A \subseteq \mathfrak{M}$ if there is a formula $\phi(v_0, \ldots, v_n)$ and a finite sequence of elements $a_1, \ldots, a_{n-1}$ in A such that
$$X = \{x \in \mathfrak{M} : \mathfrak{M} \vDash \phi[x, a_1, \ldots, a_n]\}.$$

Definition 4.3. If $A \subseteq \mathfrak{M}$ then the *algebraic closure* of A, written $\mathrm{cl}(A)$, is the union of all *finite* sets definable from A. A is said to be *algebraically closed* if $\mathrm{cl}(A) = A$. x is *algebraic* over A if $x \in \mathrm{cl}(A)$.

Lemma 4.4. *Let* $A, B \subseteq \mathfrak{M}$.

(i) $A \subseteq \mathrm{cl}(A)$,

(ii) $A \subseteq B$ *implies* $\mathrm{cl}(A) \subseteq \mathrm{cl}(B)$,

(iii) $\mathrm{cl}(\mathrm{cl}(A)) = \mathrm{cl}(A)$,

(iv) A *is algebraically closed implies* $A = \bigcup \{\mathrm{cl}(C): C \subseteq A \,\&\, C \text{ is finite}\}$,

(v) *If* $C \in \mathscr{C}$ *implies* C *is algebraically closed then* $\bigcap \mathscr{C}$ *is algebraically closed.*

Proof. (i) $v_0 = a$ defines $\{a\}$ for $a \in A$. (ii) is clear.

(iii) Because of (i) it suffices to prove that if x is algebraic over B and every element of B is algebraic over A then x is algebraic over A. x algebraic over B means there is a formula $\phi(v_0, \ldots, v_n)$ and elements $b_1, \ldots, b_n$ in B such that $x, b_1, \ldots, b_n$ satisfy ϕ and $T \vdash \exists !^k v_0 \, \phi(v_0, \ldots, v_n)$ for some natural number k.

Now for each b_i there is a formula $\psi_i(v_0)$ with parameters from A such that $(\mathfrak{M}, A) \models \psi_i(b_i)$ and $T \vdash \exists !^{k_i} v_0 \, \psi_i(v_0)$ for a least natural number k_i.

Since b_n satisfies $(\exists !^k v_0) \, \phi(v_0, b_1, \ldots, b_{n-1}, v_n)$ and k_n was chosen least possible, any element satisfying $\psi_n(v_n)$ also satisfies this same formula. So there are exact $k \, k_n$ elements satisfying

$$\exists v_n (\psi_n(v_n) \,\&\, \phi(v_0, b_1, \ldots, b_{n-1}, v_n)).$$

Eliminate $b_{n-1}, \ldots, b_1$ successively to show $x \in \mathrm{cl}(A)$.

(iv) is clear since each formula contains only a finite number of parameters.

(v) By (ii) $\mathrm{cl}(\bigcap \mathscr{C}) \subseteq \bigcap \{\mathrm{cl}(A): A \in \mathscr{C}\} = \bigcap \mathscr{C}$, since $A = \mathrm{cl}(A)$ for $A \in \mathscr{C}$. Hence by (i), $\mathrm{cl}(\bigcap \mathscr{C}) = \bigcap \mathscr{C}$. $\quad\square$

Example 4.1. Let $T(\mathbb{S})$ be the theory of an infinite set (with equality). Then the atoms of $B_n(T(\mathbb{S}))$ are the formulae which say exactly which of $v_0, \ldots, v_{n-1}$ are equal and which are distinct. The algebraic closure of a set is just itself since, given a finite subset A of the infinite set $\mathfrak{M}$, the only finite sets definable from A are the (finite) subsets of A. (Exercise.)

Example 4.2. Let $T(\mathbb{L})$ be the theory of a dense unbordered linear ordering. Then the atoms of $B_n(T(\mathbb{L}))$ are the formulae which exactly specify the finite ordering of $v_0, \ldots, v_{n-1}$ and which are equal. Again every subset is algebraically closed since if x is not in the finite set A then either x comes before or after every element of A (and there are infinitely many such elements as the ordering is unbordered) or x comes between two distinct elements of A (and again there are infinitely many such as the ordering is dense).

Example 4.3. Let $T(\mathbb{B})$ be the theory of an atomless Boolean algebra. Then a subset A of a model $\mathfrak{M}$ of $T(\mathbb{B})$ is algebraically closed if and only if A is a subalgebra of $\mathfrak{M}$. (Exercise.) Note that this does not depend on the particular language used for the axiomatization of $T(\mathbb{B})$. Also the algebraic closure of a finite set is finite, namely, it is the finite Boolean subalgebra generated by the given set.

Example 4.4. Let $T(\mathbb{V})$ be the theory of an infinite dimensional vector space over a finite field K. A subset A of a model $\mathfrak{M}$ of $T(\mathbb{V})$ is algebraically closed if and only if A is a subspace of $\mathfrak{M}$. The algebraic closure of a finite set is finite if, and only if, A is finite and in this case it is a finite dimensional vector subspace.

5. Copying

As usual a *substructure* $\mathfrak{N}$ of an L-structure $\mathfrak{M}$ is a subset of $\mathfrak{M}$ containing the constants of $\mathfrak{M}$ which has as its relations those of $\mathfrak{M}$ restricted to $|\mathfrak{N}|$ and as its constants those of $\mathfrak{M}$. We write $M(T)$ for the models of T and $N(T)$ for the substructures of models of T.

Definition 5.1. Let $\mathfrak{A}, \mathfrak{B} \in N(T)$, where $\mathfrak{A} \subseteq \mathfrak{M} \in M(T)$ and $\mathfrak{B} \subseteq \mathfrak{M}' \in M(T)$, then an injection $p: A \to B$ is said to be an *elementary monomorphism* if $(\mathfrak{M}, a)_{a \in A} \equiv (\mathfrak{M}', p(a))_{a \in A}$.

Thus p is an elementary monomorphism if and only if for all finite sequences $a_0, \ldots, a_{n-1}$ of elements of A and all formulae $\phi(v_0, \ldots, v_{n-1})$ of L:

(1) $\mathfrak{A} \models \phi[a_0, \ldots, a_{n-1}]$ if $\mathfrak{B} \models \phi[p(a_0), \ldots, p(a_{n-1})]$.

If $A = \{a_0, \ldots, a_{n-1}\}$ is finite and satisfies an atom then this condition simplifies. Then p is an elementary monomorphism if, and only if, (1) holds for the atom ϕ of $B_n(T)$ which $a_0, \ldots, a_{n-1}$ satisfy.

Lemma 5.2. Let $\mathfrak{M}, \mathfrak{M}'$ be atomic models of T, $A \subseteq \mathfrak{M}$, $B \subseteq \mathfrak{M}'$ and $p: A \to B$ an elementary monomorphism. Then p can be extended to an elementary monomorphism $\mathrm{cl}(p)$ such that the following diagram commutes (where the vertical arrows are inclusions).

$$\mathrm{cl}(A) \xrightarrow{\ \mathrm{cl}(p)\ } \mathrm{cl}(B)$$

$$\uparrow \qquad\qquad \uparrow$$

$$A \xrightarrow[\ p\ ]{} B$$

Proof. Let $p: A \to B$ be an elementary monomorphism then by Lemma 4.4(iv) it suffices to show that p can be extended to one extra point $a \in \mathrm{cl}(A) - A$ where $p(a) \in \mathrm{cl}(B)$, for then the result will follow by induction. Let $a \in \mathrm{cl}(A) - A$ then a is a solution in $\mathfrak{M}$ of a formula ψ with parameters from A. Let $\psi(a_0, a_1, \ldots, a_{n-1}, v_n)$ be chosen so that k is the least possible number such that for any ψ,

$$\mathfrak{M} \models \exists!^k v_n \, \psi(a_0, a_1, \ldots, a_{n-1}, v_n).$$

Since $\mathfrak{M}$ is atomic we may assume ψ is an atom. Let ϕ be the atom of $B_n(T)$ satisfied by $a_0, \ldots, a_{n-1}$. Then

$$\mathfrak{M} \models \phi(v_0, \ldots, v_{n-1}) \to \exists!^k v_n \, \psi(v_0, \ldots, v_n).$$

Since a satisfies $\psi(a_0, \ldots, a_{n-1}, v_n)$ in $\mathfrak{M}$ and p is not defined on a there are $\leq k-1$ elements $a' \in A$ such that $\psi(a_0, \ldots, a_{n-1}, a')$ is true in $\mathfrak{M}$. Now p is an elementary monomorphism so there are $\leq k-1$ elements b' in the range of p such that $\psi(p\,a_0, \ldots, p\,a_{n-1}, b')$ is true in $\mathfrak{M}'$. But $\phi(a_0, \ldots, a_{n-1})$ being true in $\mathfrak{M}$ implies $\phi(p\,a_0, \ldots, p\,a_{n-1})$ is true in $\mathfrak{M}'$ so $\exists!^k v_n \, \psi(p\,a_0, \ldots, p\,a_{n-1}, v_n)$ is true in $\mathfrak{M}'$. Therefore there is at least one b in $|\mathfrak{M}'| - \mathrm{ran}\, p$ such that $\psi(p\,a_0, \ldots, p\,a_{n-1}, b)$ is true in $\mathfrak{M}'$. Since $p\,a_0, \ldots, p\,a_{n-1} \in B$, b is in $\mathrm{cl}\, B$. But then $p \cup \{(a, b)\}$ is the desired extension of p.

Now we show that this extension is an elementary monomorphism. For let $a'_0, \ldots, a'_r \in A$ and suppose $\chi(a'_0, \ldots, a'_r, a)$ is true in $\mathfrak{M}$. By the the choice of k and ψ

$$\forall v_n \big(\psi(a_0, \ldots, a_{n-1}, v_n) \to \chi(a'_0, \ldots, a'_r, v_n) \big)$$

is true in $\mathfrak{M}$. Since p is elementary it follows that

$$\forall v_n \big(\psi(p\,a_0, \ldots, p\,a_{n-1}, v_n) \to \chi(p\,a'_0, \ldots, p\,a'_r, v_n) \big)$$

is true in $\mathfrak{M}'$. But $\psi(p\,a_0, \ldots, p\,a_{n-1}, b)$ is true in $\mathfrak{M}'$, so $\chi(p\,a'_0, \ldots, p\,a'_r, b)$ is true in $\mathfrak{M}'$ as was to be shown.

If p is a surjection, then clearly $\mathrm{cl}(p)$ is also a surjection. But for any p, if p' is p with codomain restricted to $p(A)$, then p' is surjective. Hence the diagram commutes. $\square$

Corollary 5.3. *Let $\mathfrak{M}, \mathfrak{M}'$ be atomic models of T, $A \subseteq \mathfrak{M}$, $B \subseteq \mathfrak{M}'$ and $p: A \to B$ an elementary isomorphism. If A is algebraically closed, B is algebraically closed.*

Lemma 5.4. *Let $\mathfrak{M}$ be an infinite atomic model. Suppose $A \subset B \subseteq |\mathfrak{M}|$, A, B are algebraically closed and B is finite. Then there is an infinite collection $\mathscr{B}$ of algebraically closed sets such that*

(i) *any two distinct members of $\mathscr{B}$ have intersection A,*

(ii) *the identity on A can be extended to an elementary isomorphism from B to any $B' \in \mathscr{B}$,*

(iii) *if N is a finite set disjoint from A then there are infinitely many B' in $\mathscr{B}$ with $N \cap B' = \emptyset$.*

Proof. Say that C *covers* A if C is algebraically closed and there is no algebraically closed set D with $A \subset D \subset C$. Since B is finite and $A \subset B$, there is a maximal chain of algebraically closed sets $A = A_0 \subset A_1 \subset A_2 \cdots \subset A_n = B$ such that A_m covers A_{m-1}. It therefore suffices to prove the theorem under the additional hypothesis that B covers A, for the general case then follows by induction on the length of a maximal chain from A to B. Let $A = \{a_0, \ldots, a_m\}$ and $B - A = \{b_0, \ldots, b_n\}$. Let $\psi(v_0, \ldots, v_{m+1})$ be the atom satisfied by $a_0, \ldots, a_m, b_0$. Since $b_0 \notin A$ and A is algebraically closed there are infinitely many elements b_0^i $(i = 0, 1, 2, \ldots)$ such that $a_0, \ldots, a_m, b_0^i$ satisfy ψ. (*) Let $\phi(v_0, \ldots, v_{m+n+1})$ be the atom satisfied by $a_0, \ldots, a_m, b_0, \ldots, b_n$ then

$$T \vdash \psi(v_0, \ldots, v_{m+1}) \to \exists v_{m+2} \ldots \exists v_{n+m+1} \phi(v_0, \ldots, v_{m+n+1})$$

and

$$T \vdash \phi(v_0, \ldots, v_{m+n+1}) \to \psi(v_0, \ldots, v_{m+1}).$$

But $a_0, \ldots, a_m, b_0^i$ satisfy ψ so there are $b_1^i, \ldots, b_n^i$ such that $a_0, \ldots, a_m, b_0^i, \ldots, b_n^i$ satisfy ϕ. Then $B^i = \{a_0, \ldots, a_m, b_0^i, \ldots, b_n^i\}$ is algebraically closed, since any $(m+n+2)$-tuple satisfying ϕ is algebraically closed. Let $p_i(a_i) = a_i$, $p_i(b_j) = b_j^i$ then p_i is an elementary monomorphism of B onto B^i by the remark preceding Lemma 5.2. Since B covers A, B^i covers A also. Hence $B^i \neq B^j$ implies $B^i \cap B^j = A$.

In order to complete the proof we show that the b_0^i can be chosen so that conditions (ii), (iii) are satisfied. We modify the choice of the b_0^i at the stage marked (*) above. Set $C^i = \mathrm{cl}\{a_0, \ldots, a_m, b_0^i\}$ with b_0^i as above. Let $D^0 = C^0$ and let D^j be the first C^i such that $C^i \neq C^k$ for any $k < j$. Then by the proof above $D^i \cap D^j = A$ if $i \neq j$ so condition (ii) is satisfied by any infinite collection of the D^i. Now $D^i \cap D^j \cap N = A \cap N = \emptyset$ if $i \neq j$ so the elements of N are distributed among a finite number of the D^i. Let the B^i be those D^i which contain no member of N, then condition (iii) is also satisfied. $\square$

We now have two very important copying lemmata which will be most useful in the sequel. The first, the Duplication Lemma, says we can copy a structure sitting on top of a given finite structure as many times as we like avoiding a finite hazard; and the second, the Endomorphism Lemma, that we can copy the whole of $\mathfrak{M}$ into itself, again avoiding a finite hazard.

Lemma 5.5 (The Duplication Lemma). *Let $\mathfrak{M}$ be an atomic model. Let A, B be finite algebraically closed subsets of $\mathfrak{M}$ and $p: A \to B$ an elementary isomorphism. Let N be a finite subset of $\mathfrak{M}$ such that $N \cap B = \emptyset$. If A' is a finite algebraically closed set containing A then there is an* *algebraically closed set* $B' \supseteq B$ *such that* $B' \cap N = \emptyset$ *and an elementary isomorphism* p' *extending* p *such that* $p': A' \cong B'$. *Moreover there are infinitely many such* B' *such that the intersection of any two distinct ones is* B.

Proof. By the preceding lemma it suffices to find a $B' \supseteq B$ and an elementary isomorphism p' extending p such that $p': A' \cong B'$.

Let $a_0, \dots, a_m$ be an enumeration of A and $a'_0, \dots, a'_n$ an enumeration of $A' - A$. Let $\psi(v_0, \dots, v_{m+n+1})$ be the atom satisfied by $a_0, \dots, a_m$, $a'_0, \dots, a'_n$ and let $\phi(v_0, \dots, v_m)$ be the atom satisfied by $a_0, \dots, a_m$. Then

$$T \vdash \phi(v_0, \dots, v_m) \to \exists v_{m+1} \dots \exists v_{m+n+1} \, \psi(v_0, \dots, v_{m+n+1})$$

since $\phi(v_0, \dots, v_m)$ and $\psi(v_0, \dots, v_{m+n+1})$ are consistent with T and T is complete. Since $p(a_0), \dots, p(a_m)$ satisfy ϕ there exist $b'_0, \dots, b'_n$ such that $p(a_0), \dots, p(a_m)$, $b'_0, \dots, b'_n$ satisfy ψ. Set $p'(a_i) = p(a_i)$, $p'(a'_j) = b'_j$ and $B' = \{pa_0, \dots, pa_n, b'_0, \dots, b'_n\}$ then p' extends p and maps A' onto B'. Moreover, by the remark preceding Lemma 5.2 p' is an elementary isomorphism. This completes the proof. ☐

Lemma 5.6 (The Endomorphism Lemma). *Let* $\mathfrak{M}$ *be a countably infinite atomic model such that the algebraic closure of a finite set is finite. Let* A, B *be finite algebraically closed subsets of* $\mathfrak{M}$ *and* $p: A \to B$ *an elementary isomorphism. Let* N *be a finite subset of* $\mathfrak{M}$ *such that* $N \cap B = \emptyset$. *Then there is an elementary monomorphism* p' *from* $\mathfrak{M}$ *into* $\mathfrak{M}$ *such that* p' *extends* p *and* $p'(\mathfrak{M}) \cap N = \emptyset$.

Proof. Since $\mathfrak{M}$ is countable let $m_0, m_1, \dots$ be an enumeration of $\mathfrak{M}$ beginning with the elements of A. Let $A_0 = A$ and for $i > 0$, let $A_i = \mathrm{cl}\{m_0, \dots, m_{i-1}\}$, then $A_0 \subseteq A_1 \subseteq \cdots$ and $\bigcup A_i = \mathfrak{M}$. Now set $p_0 = p$ and then using the Duplication Lemma recursively define elementary isomorphisms p_i with domain A_i such that p_i extends p_j for $j < i$ and $p_i(A_i) \cap N = \emptyset$. Then $p' = \bigcup p_i$ is the required isomorphism. ☐

Definition 5.7. A theory T is said to be $\aleph_0$-*categorical* if all countable models of T are isomorphic. A structure $\mathfrak{M}$ is said to be $\aleph_0$-*categorical* if it is countably infinite and its theory is $\aleph_0$-categorical.

The following theorem, due to Ryll-Nardzewski, is well-known and a proof may be found in Vaught [1959].

Theorem 5.8. *Let* T *be a complete theory with an infinite model.* T *is* $\aleph_0$-*categorical if and only if for each* n *the Boolean algebra* $B_n(T)$ *of formulae with* n *free variables is finite.* ☐

Since finite Boolean algebras are atomistic any countable model of an $\aleph_0$-categorical theory is atomic by Theorem 4.2.

Lemma 5.9. *Let $\mathfrak{M}$ be an $\aleph_0$-categorical model and A be a finite subset of $\mathfrak{M}$. Then $\mathrm{cl}(A)$ is finite. Moreover there is a function $f: \omega \to \omega$ such that $\mathrm{card}(A) \le n$ implies $\mathrm{card}(\mathrm{cl}(A)) \le f(n)$.*

Proof. Suppose card $A = n$ and $A = \{a_0, \ldots, a_{n-1}\}$ then $a_0, \ldots, a_{n-1}$ satisfy an atom $\phi(v_0, \ldots, v_{n-1})$ of $B_n(T)$. If $x \in \mathrm{cl}(A)$ then $a_0, \ldots, a_{n-1}, x$ satisfy an atom $\psi(v_0, \ldots, v_{n-1}, v_n)$ of $B_{n+1}(T)$. Since ψ is consistent with ϕ, $T \vdash \psi \to \phi$ and since x is algebraic over A there exists $k = k(\phi, \psi)$ such that $T \vdash \phi \to \exists!^k v_n \psi(v_0, \ldots, v_n)$. Since $B_n(T)$ and $B_{n+1}(T)$ are finite the sum of all $k(\phi, \psi)$ with $\phi \in B_n(T)$, $\psi \in B_{n+1}(T)$ is finite and equals $f(n)$, say. Clearly $\mathrm{card}(\mathrm{cl}(A)) \le f(n)$. $\square$

6. Dimension

In this section we present a simple version of Marsh's elegant [1966] theory of dimension which exploits familiar notions of vector space theory in certain models. Marsh uses a definition of "minimal" formula which is weaker than ours and a definition of strongly minimal formula which is stronger than our minimal. This strongly minimal formula is the same as a minimal generator (Sacks [1972]) or a point of rank 1, degree 1 in the Stone space of $B_1(T)$ (Morley [1965]).

Definition 6.1. A formula $\phi(v_0)$ in the language L is said to be *minimal* in $\mathfrak{M}$ if (i) $\phi(v_0)$ is satisfied by infinitely many elements of $\mathfrak{M}$ and (ii) for any formula $\psi(v_0) = \psi(v_0, a_1, \ldots, a_n)$ with parameters $a_1, \ldots, a_n$ from $\mathfrak{M}$ then one of

$$\phi(v_0) \& \psi(v_0), \qquad \phi(v_0) \& \neg \psi(v_0)$$

is satisfied by only finitely many elements of $\mathfrak{M}$.

Lemma 6.2 (The Exchange Lemma). *Let X be a subset of $\mathfrak{M}$ and a, b be elements not algebraic over X. If a satisfies a minimal formula in $\mathfrak{M}$ and $b \in \mathrm{cl}(X \cup \{a\})$ then $a \in \mathrm{cl}(X \cup \{b\})$.*

Proof. As usual let $M = |\mathfrak{M}|$ and $T(M)$ be the theory of $\mathfrak{M}$ in an extended language with constants for the elements of M. If $b \in \mathrm{cl}(X \cup \{a\})$ then there is a formula $\psi(v_0, v_1)$ with parameters from X such that

(i) $T(M) \vdash \psi(a, b)$, and

(ii) for some natural number k, $T(M) \vdash \exists!^k v_1 \psi(a, v_1)$.

If $\psi(v_0, b) \& \phi(v_0)$ is only satisfied by a finite number of elements of $\mathfrak{M}$, then a is algebraic over $X \cup \{b\}$ since a is such an element. Suppose not, then there are infinitely many elements satisfying $\psi(v_0, b) \& \phi(v_0)$. Since $\phi(v_0)$ is minimal $\neg \psi(v_0, b) \& \phi(v_0)$ is satisfied by only finitely many, say l, elements of $\mathfrak{M}$. Let $\chi(v_1)$ be the formula

$$\exists !^l \, v_0 \big(\neg \psi(v_0, v_1) \& \phi(v_0) \big).$$

Then $T(\mathfrak{M}) \vdash \chi(b)$. Since $b \notin \mathrm{cl}(X)$ and χ has parameters only from X, $\chi(v_1)$ is satisfied by infinitely many elements of $\mathfrak{M}$. Let $b_0, \ldots, b_k$ be $k+1$ distinct such elements. For $i = 0, \ldots, k$ let S^i be the set of l elements of $\mathfrak{M}$ satisfying $\neg \psi(v_0, b_i) \& \phi(v_0)$. Now, if $x \in \mathfrak{M} - S^i$ and x satisfies $\phi(v_0)$, then x is not one of the l elements satisfying $\neg \psi(v_0, b_i)$ so x satisfies $\psi(v_0, b_i)$. Hence if $x \in \mathfrak{M} - \bigcup S^i$ and x satisfies $\phi(v_0)$ then x satisfies

$$\psi(v_0, b_0) \& \ldots \& \psi(v_0, b_k).$$

Therefore x does not satisfy

(1) $$\exists !^k \, v_1 \big(\psi(v_0, v_1) \& \phi(v_0) \big).$$

It follows that any element of $\mathfrak{M}$ which does satisfy (1) is in $\bigcup S^i$, which is a $(k+1)l$ element set. But then, since (1) only contains parameters from X, any member of $\bigcup S^i$ is in $\mathrm{cl}(X)$. Now by (ii) a satisfies (1) so $a \in \mathrm{cl}(X)$ contrary to hypothesis. $\square$

Say that an element a of $\mathfrak{M}$ is a *solution* of a formula $\chi(v_0, b_1, \ldots, b_n)$ where $b_1, \ldots, b_n \in \mathfrak{M}$ if a satisfies χ. If $\psi(v_0)$ has only finitely many solutions and ψ has no parameters then any solution of ψ is algebraic over the empty set and we shall simply say such an element is *algebraic*.

Let $\mathrm{Min}(\mathfrak{M})$ be the set of all solutions of minimal formulae in $\mathfrak{M}$.

Definition 6.3. If $X \subseteq \mathfrak{M}$ then X is said to be *independent* if, for all $x \in X$, $x \notin \mathrm{cl}(X - \{x\})$.

Clearly a subset of an independent set is independent. Also X is independent if and only if every finite subset of X is independent.

We now proceed as for vector spaces and, as most of the proofs are literally identical with the corresponding vector space proofs, we omit them (see for example van der Waerden [1931] Chapter V).

Lemma 6.4. (ia) *If $\{a_0, \ldots, a_{n-1}\}$ is independent but $\{a_0, \ldots, a_n\}$ is not, then for some i*

$$a_i \in \mathrm{cl}\{a_0, \ldots, a_{i-1}, a_{i+1}, \ldots, a_n\}.$$

(ib) *If $\{a_0, \ldots, a_{n-1}\}$ is independent but $\{a_0, \ldots, a_n\}$ is not and a_n satisfies a minimal formula in $\mathfrak{M}$ then*

$$a_n \in \mathrm{cl}\{a_0, \ldots, a_{n-1}\}.$$

(ii) *Suppose $X, Y \subseteq \mathrm{Min}(\mathfrak{M})$, $\mathrm{cl}(X) \subseteq \mathrm{cl}(Y)$ and X is independent then (a) $\mathrm{card}(X) \leq \mathrm{card}(Y)$ and (b) there is a subset Y_0 of Y such that $X \cup Y_0$ is independent and $\mathrm{cl}(X \cup Y_0) = \mathrm{cl}(Y)$.* $\square$

We remark that Lemma 6.4 can be extended into the transfinite by trivial modifications but we do not need this here.

Definition 6.5. Let $Y = \mathrm{cl}(X)$ where $X \subseteq \mathrm{Min}(\mathfrak{M})$ then the *dimension* of Y written $\dim(Y)$, is the number of elements in any independent set $Y_0 \subseteq \mathrm{Min}(\mathfrak{M})$ such that $\mathrm{cl}(Y_0) = Y$. Such a Y_0 is said to be a *basis* for Y.

By Lemma 6.4, dimension is well-defined.

Example 6.1. For $\mathfrak{M}$ the $\aleph_0$-categorical model for the theory of equality with an infinite number of elements, a minimal formula is $v_0 = v_0$, $\mathfrak{M} = \mathrm{Min}(\mathfrak{M})$ and $\dim(X) = \mathrm{card}(X)$.

Example 6.2. For $\mathfrak{M}$ the $\aleph_0$-catecorical model for the theory of an infinite dimensional vector space over a fixed finite field a minimal formula is $v_0 = v_0$, $\mathfrak{M} = \mathrm{Min}(\mathfrak{M})$ and dimension is the familiar vector space dimension.

If Y is algebraically closed we say that X spans Y if $Y \subseteq \mathrm{cl}(X)$. Thus if $X \subseteq \mathrm{Min}(\mathfrak{M})$ then X is a basis for Y if, and only if, X is a maximal independent subset of Y if, and only if, X is a minimal subset which spans Y.

Corollary 6.6. *Let X, Y be algebraically closed subsets of $\mathrm{cl}(\mathrm{Min}(\mathfrak{M}))$.*

(i) *X has a basis,*

(ii) *all bases for X have the same number of elements,*

(iii) *if $X \subset Y$ then $1 + \dim X \le \dim Y$.* $\square$

Corollary 6.7. *Suppose $X_1 \cup X_2$ is independent and $X_1 \cup X_2 \subseteq \mathrm{Min}(\mathfrak{M})$. Then*

$$\mathrm{cl}(X_1 \cap X_2) = \mathrm{cl}(X_1) \cap \mathrm{cl}(X_2).$$

Proof. It suffices to consider X_1, X_2 finite. For the significant direction suppose $a \in \mathrm{cl}(X_1) \cap \mathrm{cl}(X_2)$ but $a \notin \mathrm{cl}(X_1 \cap X_2)$. Then there exist $x_i \in X_i - (X_1 \cap X_2)$. By the Exchange Lemma, $x_i \in \mathrm{cl}(X_i - \{x_i\}) \cup \{a\})$ so

$$X_1 \cup X_2 \subseteq \mathrm{cl}\big((X_1 - \{x_1\}) \cup (X_2 - \{x_2\}) \cup \{a\}\big).$$

But then

$$\dim \mathrm{cl}(X_1 \cup X_2) \le (\dim \mathrm{cl}(X_1) - 1) + (\dim \mathrm{cl}(X_2) - 1) + 1.$$

But $X_1 \cup X_2$ is independent, so

$$\dim \mathrm{cl}(X_1 \cup X_2) = \dim \mathrm{cl}\, X_1 + \dim \mathrm{cl}\, X_2$$

which is a contradiction. $\square$

By virtue of Corollary 6.7, if X is independent, $X \subseteq \mathrm{Min}(\mathfrak{M})$ and $a \in \mathrm{cl}(X)$ then there is a smallest (finite) set $X' \subseteq X$ such that $a \in \mathrm{cl}\, X'$. We call X' the *support* of a and write $X' = \mathrm{supp}(a)$. Then for $Y \subseteq X$, $a \in \mathrm{cl}(Y)$ if, and only if, $\mathrm{supp}(a) \subseteq Y$.

Corollary 6.8. *If* $\{S^i: i\in I\}\subseteq\mathscr{S}$ *is a set of algebraically closed sets with* $S^i=\mathrm{cl}(X^i)$ *where* $X^i\subseteq\mathrm{Min}(\mathfrak{M})$ *and* $\mathscr{S}$ *is a chain under inclusion then* $\dim\cup\mathscr{S}=\sup\{\dim S^i: i\in I\}$.

Finally we prove strengthened versions of the Duplication and Endomorphism Lemmata.

We say that a structure $\mathfrak{M}$ *has dimension* if there is a minimal formula ϕ satisfied by all non-algebraic elements of $\mathfrak{M}$. In this case the dimension function assigns to each subset X of $\mathfrak{M}$ the cardinal of a basis for X. (In this case also, ϕ is essentially unique for if ψ is any other minimal formula satisfied by all non-algebraic elements of $\mathfrak{M}$ then either the solutions of ϕ and ψ differ only by a finite set or those of ϕ and $\neg\psi$ differ only by a finite set. Thus a minimal formula is $v_0=v_0$.)

Lemma 6.9. *Let* $\mathfrak{M}$, $\mathfrak{M}'$ *be models of* T *and let* $\phi(v_0)$ *be a minimal formula in* $\mathfrak{M}$. *Let* $A\subseteq\mathfrak{M}$ *and* $B\subseteq\mathfrak{M}'$ *be independent sets consisting of solutions of* $\phi(v_0)$. *Let* $p: A\to B$ *be a one-one map then* p *is an elementary monomorphism.*

Proof. It suffices to consider the case when A is a finite set. We now establish the result by induction on the cardinal of A. Suppose it holds for n and $A=\{a_0,\dots,a_{n-1},a_n\}$. So assume p restricted to $a_0,\dots,a_{n-1}$ is an elementary monomorphism and let $\psi(v_0,\dots,v_n)$ be any formula satisfied by $a_0,a_1,\dots,a_n$. Then a_n satisfies

(2) $$\psi(a_0,\dots,a_{n-1},v_n)\,\&\,\phi(v_n).$$

Since $a_n\notin\mathrm{cl}\{a_0,\dots,a_{n-1}\}$ the formula (2) has infinitely many solutions in $\mathfrak{M}$. Since ϕ is minimal there is a natural number k such that in $\mathfrak{M}$

(3) $$\exists^{\leq k}v_n(\neg\psi(v_0,\dots,v_{n-1},v_n)\,\&\,\phi(v_n))$$

is satisfied by $a_0,\dots,a_{n-1}$. Since p restricted to $a_0,\dots,a_{n-1}$ is an elementary monomorphism, $p(a_0),\dots,p(a_{n-1})$ satisfy (3) in $\mathfrak{M}'$. Hence if $p(a_i)=b_i$,

(4) $$\neg\psi(b_0,\dots,b_{n-1},v_n)\,\&\,\phi(v_n)$$

has only finitely many solutions in $\mathfrak{M}'$. Now $p(a_n)\notin\mathrm{cl}\{p(a_0),\dots,p(a_{n-1})\}$, so $p(a_n)$ is not a solution of (4) in $\mathfrak{M}'$. Since $p(a_n)$ is a solution of $\phi(v_n)$, it follows that $\psi(b_0,\dots,b_n)$ holds in $\mathfrak{M}'$. $\square$

Lemma 6.10 (Duplication Lemma revised). *Assume the hypothesis of the Duplication Lemma 5.5 and that* $\mathfrak{M}$ *has dimension. Then* $\mathscr{B}=\{B^0,B^1,\dots\}$ *can be chosen so that for all* i, $B_i\cap\mathrm{cl}(\cup\{B^i: j\neq i\})=B$ *and* $\mathrm{cl}(\cup B^j)\cap N=\emptyset$.

Proof. Let $\dim A=m=\dim B$ and $\dim A'=m+n$ ($n\geq0$ since $A'\supseteq A$). Let $b_0,\dots,b_{m-1}$ be a basis for B. Now we can find by induction

$b_m, b_{m+1}, \ldots$ in $\mathfrak{M}$ such that $b_i \notin \mathrm{cl}\{b_0, \ldots, b_{m-1}, b_m, \ldots, b_{i-1}\}$ and $\mathrm{cl}\{b_0, \ldots, b_i\} \cap N = \emptyset$. By the preceding lemma

$$B_0 = \mathrm{cl}\{b_0, \ldots, b_{m-1}, b_m, \ldots, b_{m+n-1}\},$$
$$B_1 = \mathrm{cl}\{b_0, \ldots, b_{m-1}, b_{m+n}, \ldots, b_{m+2n-1}\},$$

and in general

$$B_i = \mathrm{cl}\{b_0, \ldots, b_{m-1}, b_{m+ni}, \ldots, b_{m+ni-1}\}$$

then have the required properties. $\square$

Lemma 6.11 (Endomorphism Lemma revised). *Assume the hypothesis of the Endomorphism Lemma 5.6 and that $\mathfrak{M}$ has dimension. Then the extension p' can be chosen so that $\mathrm{cl}(\mathrm{ran}\, p') \cap N = \emptyset$.*

Proof. Similar to that for Lemma 6.10 and therefore left to the reader. $\square$

Part III. Combinatorial Functions

7. Strict Combinatorial Functors

In an important special case a particular class of combinatorial functors, the strict combinatorial functors, play a most important role. This section describes the special case and establishes the basic properties of strict functors; such as existence, extent of uniqueness and closure under composition.

Throughout this section all models $\mathfrak{M}$ are countably infinite atomic models in which the algebraic closure of a finite set is finite. $\mathbb{C}(\mathfrak{M})$ is the category whose objects are the algebraically closed subsets of $\mathfrak{M}$, whose morphisms are the elementary monomorphisms between algebraically closed sets. $^{\circ}\mathbb{C}(\mathfrak{M})$ is the full subcategory of finite objects and morphisms. Such $\mathbb{C}$ are appropriate in the sense of Section 1 (Exercise).

We say that $\mathfrak{M}$ has *degree* 1 for any independent $a_1, \ldots, a_n$ in $\mathfrak{M}$, if $a \in \mathrm{cl}\{a_1, \ldots, a_n\}$, then there is a formula $\phi(v_0, \ldots, v_n)$ such that a is the unique solution in $\mathfrak{M}$ to $\phi(v_0, a_1, \ldots, a_n)$.

Example 7.1. The integers under equality are degree 1 because $a \in \mathrm{cl}\{a_1, \ldots, a_n\}$ if, and only if, $a = a_i$ for some i. So $v_0 = a_i$ is the desired formula.

Example 7.2. A vector space over a field, with constants for scalars, is degree 1 because $a \in \mathrm{cl}\{a_1, \ldots, a_n\}$ if, and only if, $a = \lambda_1 a_1 + \cdots + \lambda_n a_n$ for suitable scalars $\lambda_1, \ldots, \lambda_n$. So $v_0 = \lambda_1 a_1 + \cdots + \lambda_n a_n$ is the desired formula.

The main simplifying point of assuming $\mathfrak{M}$ is degree 1 is that if A, B are independent subsets of $\mathfrak{M}$ and $p: A \to B$ is injective, there is at most one elementary monomorphism $\mathrm{cl}\, p: \mathrm{cl}\, A \to \mathrm{cl}\, B$ extending p. So if A, B, C are independent sets satisfying the same minimal formula, $p: A \to B$, $q: B \to C$ are injections, and $\mathfrak{M}$ has degree 1, then $(\mathrm{cl}\, q) \circ (\mathrm{cl}\, p)$ and $\mathrm{cl}(q \circ p)$ must coincide. (Apply Lemma 5.2 and Corollary 6.9.) We write $\mathscr{P}(A)$, $\mathscr{P}(\mathfrak{M})$ for the set of all subsets of A, $|\mathfrak{M}|$, respectively.

Definition 7.1. A map $G: \mathrm{Ob}\big({}^\circ\mathbb{C}(\mathfrak{M})\big) \to \mathscr{P}(\mathfrak{M}')$ is said to be a *pre-combinatorial operator* if

1) $\mathfrak{A} \neq \mathfrak{A}'$ implies $G(\mathfrak{A}) \cap G(\mathfrak{A}') = \emptyset$ for $\mathfrak{A}$, $\mathfrak{A}'$ in ${}^\circ\mathbb{C}(\mathfrak{M})$,

2) $p: \mathfrak{A} \to \mathfrak{A}'$ an isomorphism of ${}^\circ\mathbb{C}(\mathfrak{M})$ implies card $G(\mathfrak{A}) =$ card $G(\mathfrak{A}')$ and

3) $\bigcup\{G(\mathfrak{A}): \mathfrak{A} \in {}^\circ\mathbb{C}\}$ is independent.

Now objects $\mathfrak{A}$, $\mathfrak{A}'$ of $\mathbb{C}(\mathfrak{M})$ have the same isomorphism type if there is an isomorphism $p: \mathfrak{A} \to \mathfrak{A}'$ in $\mathbb{C}(\mathfrak{M})$. Finite isomorphism types are isomorphism types of objects in ${}^\circ\mathbb{C}(\mathfrak{M})$. The isomorphism type $\langle \mathfrak{A} \rangle$ of a finite object $\mathfrak{A}$ consists of all $\mathfrak{A}' \in \mathbb{C}$ such that $\mathfrak{A}'$ is of the same isomorphism type as $\mathfrak{A}$.

Lemma 7.2. *Each precombinatorial operator* $G: \mathrm{Ob}\,{}^\circ\mathbb{C}(\mathfrak{M}) \to \mathscr{P}(\mathfrak{M}')$ *induces a map g on finite isomorphism types from $\mathbb{C}$ to*

$$E^+ = \{0, 1, 2, \ldots\} \cup \{\aleph_0\}$$

given by $g(\langle \mathfrak{A} \rangle) = \mathrm{card}\, G\mathfrak{A}$. *Conversely if $\mathfrak{M}'$ has dimension, every map d to E^+ from finite isomorphism types from $\mathbb{C}$ arises from a precombinatorial operator.*

Proof. Since $\mathfrak{M}'$ is infinite and has dimension, and the algebraic closure of a finite set is finite, $\mathfrak{M}'$ contains an infinite independent set U. Since ${}^\circ\mathbb{C}$ is countable, there is an injection $j: {}^\circ\mathbb{C} \times E \to U$. Define $G(\mathfrak{A}) = \{j(\mathfrak{A}, n): 0 \leq n < d(\langle \mathfrak{A} \rangle)\}$ for $\mathfrak{A} \in {}^\circ\mathbb{C}$. Requirement 1) holds by the choice of j, 2) by Definition 7.1 of G, and 3) by choice of U. $\square$

We call the G so constructed from d the *Myhill precombinatorial operator* corresponding to d. Note that there is a continuum of G's for each d, since there is a continuum of choices of U.

Definition 7.3. A combinatorial functor $F: \mathbb{C}(\mathfrak{M}) \to \mathbb{C}(\mathfrak{M}')$ is said to be *strict* if there is a precombinatorial operator G such that for all objects $\mathfrak{A}$ of $\mathbb{C}(\mathfrak{M})$,

$(*)$ $\qquad\qquad F(\mathfrak{A}) = \mathrm{cl} \bigcup \{G(\mathfrak{A}'): \mathfrak{A}' \subseteq \mathfrak{A}\ \&\ \mathfrak{A}' \in {}^\circ\mathbb{C}\}.$

It will presently appear that all $\mathbb{S}$-valued combinatorial functors F are strict. There

$$G(\mathfrak{A}) = F(\mathfrak{A}) - \bigcup \{F(\mathfrak{A}'): \mathfrak{A}' \in {}^\circ\mathbb{C}(\mathfrak{M})\ \&\ \mathfrak{A}' \subset \mathfrak{A}\}$$

is the corresponding precombinatorial operator (*cf.* Lemma 7.7).

Theorem 7.4. *Suppose $\mathfrak{M}'$ has dimension and degree 1. Suppose $G:$ ${}^\circ\mathbb{C}(\mathfrak{M}) \to \mathscr{P}(\mathfrak{M}')$ is precombinatorial. Then there is a strict combinatorial functor $F: \mathbb{C} \to \mathbb{C}'$ satisfying $(*)$.*

Proof. We define F on ${}^\circ\mathbb{C}$ to $\mathbb{C}'$ and apply Lemma 3.10 to extend to $\mathbb{C}$. For objects $\mathfrak{A} \in {}^\circ\mathbb{C}$, define $F(\mathfrak{A})$ by $(*)$. If $x \in \mathrm{cl} \bigcup \{G(\mathfrak{A}'): \mathfrak{A}' \in {}^\circ\mathbb{C}\}$,

let supp x be the support of x relative to the independent set $\bigcup\{G(\mathfrak{A})\colon$
$\mathfrak{A}\in{}^\circ\mathbb{C}\}$. Then (since supp x is finite) $x\in F(\mathfrak{A})$ implies that there is a finite
sequence $\mathfrak{A}^1,\ldots,\mathfrak{A}^k$ of objects in ${}^\circ\mathbb{C}$, contained in $\mathfrak{A}$, such that
$(\text{supp } x)\cap G(\mathfrak{A}^i)\neq\emptyset$. The definition of support now ensures that for all
$\mathfrak{A}, x\in F(\mathfrak{A})\leftrightarrow\mathfrak{A}^1\cup\cdots\cup\mathfrak{A}^k\subseteq\mathfrak{A}$, so $F^\leftarrow x=\mathrm{cl}(\mathfrak{A}^1\cup\cdots\cup\mathfrak{A}^k)$ will do.
According to Theorem 3.12, it remains only to show how to define Fp
for morphisms p in ${}^\circ\mathbb{C}(\mathfrak{M})$ so as to be an inclusion preserving functor.

For all $G(\mathfrak{A}')$ with $\mathfrak{A}'\in{}^\circ\mathbb{C}$, simultaneously choose in advance an
enumeration without repetitions of $G(\mathfrak{A}')$. Now let $p\colon\mathfrak{A}\to\mathfrak{B}$ be in ${}^\circ\mathbb{C}$.
Restricting p gives an isomorphism of each $\mathfrak{A}'\subseteq\mathfrak{A}$ where $\mathfrak{A}'\in{}^\circ\mathbb{C}$, with
a $\mathfrak{B}'\subseteq p(\mathfrak{A})$, where $\mathfrak{B}'\in{}^\circ\mathbb{C}$, and conversely. So by requirement 2) of
Definition 7.1, $G(\mathfrak{A}')$ and $G(\mathfrak{B}')$ have the same cardinality. By 1) of
Definition 7.1, $\mathfrak{A}'\neq\mathfrak{A}''$ implies $G(\mathfrak{A}')\cap G(\mathfrak{A}'')=\emptyset$. So we can map
$\bigcup\{G(\mathfrak{A}')\colon\mathfrak{A}'\subseteq\mathfrak{A}\,\&\,\mathfrak{A}'\in{}^\circ\mathbb{C}\}$ bijectively to $\bigcup\{G(\mathfrak{B}')\colon\mathfrak{B}'\subseteq p(\mathfrak{A})\,\&\,\mathfrak{B}'\in{}^\circ\mathbb{C}\}$
by mapping $G(\mathfrak{A}')$ to $G(\mathfrak{B}')$ in order of enumeration. Let this map be f.
Then by 3) of Definition 7.1, dom f and ran f are both independent
sets. Since $\mathfrak{M}'$ has dimension and is of degree 1, there is therefore a unique
isomorphism $\mathrm{cl}\, f\colon \mathrm{cl}\,\mathrm{dom}\, f\to\mathrm{cl}\,\mathrm{codom}\, f$. Thus $\mathrm{cl}\, f\colon F\mathfrak{A}\to F(p\mathfrak{A})$ is an
isomorphism. If $i\colon F(p\mathfrak{A})\to F\mathfrak{B}$ is the inclusion morphism, then set
$Fp=i\circ\mathrm{cl}\, f$. We note that F is a functor because of the uniqueness of
$\mathrm{cl}\, f$ (cf. our initial remarks on degree 1). $\square$

Corollary 7.5. *Suppose $\mathfrak{M}$ has dimension and degree 1. Let $F\colon E\to\mathfrak{M}$
be an injection with range a basis for $\mathfrak{M}$. Let V be defined for objects $\mathfrak{A}$
of $\mathbb{S}$ by*

$$V(\mathfrak{A})=\mathrm{cl}\big(F(\mathfrak{A})\big).$$

Then V can be extended to a strict combinatorial functor $V\colon\mathbb{S}\to\mathbb{C}(\mathfrak{M})$.

Proof. The appropriate precombinatorial operator $G\colon{}^\circ\mathbb{S}\to\mathscr{P}(\mathfrak{M})$ is

$$G(\mathfrak{A})=\{F(a)\}\quad\text{if }\mathfrak{A}=\{a\},\ G(\mathfrak{A})=\emptyset\quad\text{otherwise}.\quad\square$$

Note. An example due to M. Morley shows this fails if the degree 1
hypothesis is omitted.

Theorem 7.6. *Suppose $G^1, G^2\colon\mathrm{Ob}\big({}^\circ\mathbb{C}(\mathfrak{M})\big)\to\mathscr{P}(\mathfrak{M}')$ are precombinatorial and induce the same function from finite isomorphism types of
${}^\circ\mathbb{C}(\mathfrak{M})$ to E^+. Suppose $F^1, F^2\colon\mathbb{C}(\mathfrak{M})\to\mathbb{C}(\mathfrak{M}')$ are the respective induced
strict combinatorial functors given by $(*)$. Suppose $\mathfrak{M}'$ has dimension and
degree 1. Then the restrictions of F^1, F^2 to the subcategories of inclusion
maps are naturally equivalent.*

Proof. For each $\mathfrak{A}\in{}^\circ\mathbb{C}$ simultaneously choose enumerations without
repetitions for $G^1(\mathfrak{A})$, $G^2(\mathfrak{A})$. Let τ be the bijection with domain

$\bigcup \{G^1(\mathfrak{A}): \mathfrak{A} \in {}^\circ \mathbf{C}(\mathfrak{M})\}$, codomain $\bigcup \{G^2(\mathfrak{A}): \mathfrak{A} \in {}^\circ \mathbf{C}\}$ such that τ takes the set $G^1(\mathfrak{A})$ bijectively to $G^2(\mathfrak{A})$ in order of enumeration. τ exists because $G^1(\mathfrak{A})$ and $G^2(\mathfrak{A})$ have the same cardinality, while $G^1(\mathfrak{A})$, $G^1(\mathfrak{B})$ are disjoint if $\mathfrak{A}, \mathfrak{B}$ are distinct. Since G^1, G^2 are precombinatorial operators dom τ, codom τ are independent sets. By Lemma 5.2, Lemma 6.9 and the fact that $\mathfrak{M}$ has degree 1 there is a unique isomorphism cl τ: cl dom $\tau \to$ cl codom τ extending τ. But (∗) for F^1, F^2 shows that for every object $\mathfrak{A}$ of $\mathbf{C}(\mathfrak{M})$ there is an isomorphism $\tau_\mathfrak{A}$: $F^1(\mathfrak{A}) \to F^2(\mathfrak{A})$ given by letting $\tau_\mathfrak{A}$ be the restriction of cl τ to have domain $F^1(\mathfrak{A})$ and codomain $F^2(\mathfrak{A})$. Then $\mathfrak{A} \to \tau_\mathfrak{A}$ is the desired natural equivalence, that is, for $\mathfrak{A} \subseteq \mathfrak{B}$,

$$\begin{array}{ccc} F^1(\mathfrak{B}) & \xrightarrow{\ \tau_\mathfrak{B}\ } & F^2(\mathfrak{B}) \\ \uparrow & & \uparrow \\ \\ F^1(\mathfrak{A}) & \xrightarrow[\ \tau_\mathfrak{A}\ ]{} & F^2(\mathfrak{A}) \end{array}$$

is commutative. (The unmarked arrows are inclusion maps.) ☐

We wish to prove a result having as a special case closure of strict combinatorial functors under composition. This requires two lemmata.

For F a combinatorial functor from $\mathbf{C}(\mathfrak{M})$ to $\mathbf{C}(\mathfrak{M}')$ we define F^e: $\mathrm{Ob}({}^\circ \mathbf{C}(\mathfrak{M})) \to \mathscr{P}(\mathfrak{M}')$ by $F^e(\mathfrak{A}) = F(\mathfrak{A}) - \bigcup \{F(\mathfrak{A}'): \mathfrak{A}' \subset \mathfrak{A} \,\&\, \mathfrak{A}' \in {}^\circ \mathbf{C}\}$.

Lemma 7.7. *Let* F: $\mathbf{C}(\mathfrak{M}) \to \mathbf{C}(\mathfrak{M}')$ *be any combinatorial functor. Then*

(i) $\mathfrak{A} \neq \mathfrak{B}$ *implies* $F^e(\mathfrak{A}) \cap F^e(\mathfrak{B}) = \emptyset$, *for* $\mathfrak{A}, \mathfrak{B} \in {}^\circ \mathbf{C}(\mathfrak{M})$.

(ii) *if* p: $\mathfrak{A} \to \mathfrak{B}$ *is an isomorphism of* ${}^\circ \mathbf{C}$ *then* $F p$ *is a bijection from* $F^e(\mathfrak{A})$ *to* $F^e(\mathfrak{B})$.

(iii) *For all objects* $\mathfrak{A}$ *of* $\mathbf{C}(\mathfrak{M})$, $|F(\mathfrak{A})|$ *is the disjoint union of all* $F^e(\mathfrak{A}')$ *with* $\mathfrak{A}' \subseteq \mathfrak{A}$ *and* $\mathfrak{A}' \in {}^\circ \mathbf{C}$.

Proof. For (i), $x \in F^e(\mathfrak{A}) \cap F^e(\mathfrak{B}) \to x \in F(\mathfrak{A}) \cap F(\mathfrak{B}) \to x \in F(\mathfrak{A} \cap \mathfrak{B})$ (Lemma 3.9). By the definition of F^e, $x \in F(\mathfrak{A} \cap \mathfrak{B})$ and $x \in F^e(\mathfrak{A})$ imply $\mathfrak{A} \cap \mathfrak{B} = \mathfrak{A}$, so $\mathfrak{A} \subseteq \mathfrak{B}$. Similarly $\mathfrak{B} \subseteq \mathfrak{A}$, so $\mathfrak{A} = \mathfrak{B}$. For (ii) remember F is an inclusion preserving functor. Restrictions of p match each $\mathfrak{A}' \subseteq \mathfrak{A}$, $\mathfrak{A}' \in {}^\circ \mathbf{C}$ with a $\mathfrak{B}' \subseteq \mathfrak{B}$ such that $\mathfrak{B}' \in {}^\circ \mathbf{C}$, and conversely. Therefore restrictions of $F p$ match each $F \mathfrak{A}'$ with an $F \mathfrak{B}'$ (and conversely). So $F p$ matches $F(\mathfrak{A})$ with $F(\mathfrak{B})$ and $\bigcup \{F(\mathfrak{A}'): \mathfrak{A}' \subseteq \mathfrak{A} \,\&\, \mathfrak{A}' \in {}^\circ \mathbf{C}(\mathfrak{M})\}$ with $\bigcup \{F(\mathfrak{B}'): \mathfrak{B}' \subseteq \mathfrak{B}, \mathfrak{B}' \in {}^\circ \mathbf{C}(\mathfrak{M})\}$. So $F p$ matches $F^e \mathfrak{A}$ with $F^e \mathfrak{B}$.

(Note that $x \in F^e \mathfrak{A} \leftrightarrow F^\leftarrow x = \mathfrak{A}$.) ☐

Lemma 7.8. *Suppose* p: $\mathfrak{A} \to \mathfrak{B}$ *is an isomorphism of* ${}^\circ \mathbf{C}(\mathfrak{M})$ *and* F: $\mathbf{C}(\mathfrak{M}) \to \mathbf{C}(\mathfrak{M}')$ *is a combinatorial functor. Then the isomorphism* $F p$:

$F\mathfrak{A} \to F\mathfrak{B}$ *induces a bijection from*

$$\{\mathfrak{A}' \in {}^{\circ}\mathbb{C}(\mathfrak{M}'): F^{\leftarrow}\mathfrak{A}' = \mathfrak{A}\}$$

onto

$$\{\mathfrak{B}' \in {}^{\circ}\mathbb{C}(\mathfrak{M}'): F^{\leftarrow}\mathfrak{B}' = \mathfrak{B}\}$$

given by $\mathfrak{B}' = q(\mathfrak{A}')$ *where* q *(is the isomorphism which) is* Fp *with domain and codomain restricted to* $\mathfrak{A}'$, $\mathfrak{B}'$, *respectively.*

Proof. Since Fp is an isomorphism in ${}^{\circ}\mathbb{C}(\mathfrak{M}')$, we know Fp maps $\{\mathfrak{A}' \in {}^{\circ}\mathbb{C}(\mathfrak{M}'): \mathfrak{A}' \subseteq F\mathfrak{A}\}$ bijectively onto $\{\mathfrak{B}' \in {}^{\circ}\mathbb{C}(\mathfrak{M}'): \mathfrak{B}' \subseteq F\mathfrak{B}\}$. What must be shown is that $F^{\leftarrow}\mathfrak{A}' = \mathfrak{A}$ implies $F^{\leftarrow}\mathfrak{B}' = \mathfrak{B}$, where $\mathfrak{B}' = p(\mathfrak{A}')$. But $F^{\leftarrow}\mathfrak{A}' = \mathfrak{A}$ implies there are $x_1, \ldots, x_n \in \mathfrak{A}'$ such that, if $\mathfrak{A}_i = F^{\leftarrow}x_i$, then $\mathrm{cl} \bigcup \mathfrak{A}_i = \mathfrak{A}$. Let $\mathfrak{B}_i = p\mathfrak{A}_i$. Since p is bijective and an elementary monomorphism, $\mathrm{cl} \bigcup \mathfrak{B}_i = \mathfrak{B}$. Now $F^{\leftarrow}x_i = \mathfrak{A}_i \leftrightarrow x_i \in F^e\mathfrak{A}_i$. By the preceding Lemma 7.4, Fp matches $F^e\mathfrak{A}_i$ with $F^e\mathfrak{B}_i$, so $y_i = (Fp)x_i \in F^e\mathfrak{B}_i$, so $F^{\leftarrow}y_i = \mathfrak{B}_i$. But $x_1, \ldots, x_i \in \mathfrak{A}'$ implies $y_1, \ldots, y_k \in \mathfrak{B}'$, so $F^{\leftarrow}\mathfrak{B}' \supseteq \mathrm{cl} \bigcup F^{\leftarrow}y_i = \mathrm{cl}(\bigcup \mathfrak{B}_i) = \mathfrak{B}$. But of course $\mathfrak{B}' \subseteq F\mathfrak{B}$ implies $F^{\leftarrow}\mathfrak{B}' \subseteq \mathfrak{B}$, so $F^{\leftarrow}\mathfrak{B}' = \mathfrak{B}$. $\square$

Theorem 7.9. *Let* $\mathfrak{M}$, $\mathfrak{M}'$, $\mathfrak{M}''$ *be given, and let* $\mathfrak{M}''$ *have dimension and degree 1. Let* $\mathbb{C} = \mathbb{C}(\mathfrak{M})$, $\mathbb{C}' = \mathbb{C}(\mathfrak{M}')$ *and* $\mathbb{C}'' = \mathbb{C}(\mathfrak{M}'')$. *Suppose* $F^2: \mathbb{C} \to \mathbb{C}'$ *is a combinatorial functor and* $F^1: \mathbb{C}' \to \mathbb{C}''$ *is a strict combinatorial functor. Then* $F^1 \circ F^2: \mathbb{C} \to \mathbb{C}''$ *is a strict combinatorial functor.*

Proof. By Theorem 3.2 we know $F^1 \circ F^2$ is combinatorial. We need only produce the precombinatorial operator G^3 inducing $F^1 \circ F^2$ from the precombinatorial operator G^1 inducing F^1.

For $\mathfrak{A} \in {}^{\circ}\mathbb{C}$, define

$$G^3(\mathfrak{A}) = \bigcup \{G^1(\mathfrak{A}'): \mathfrak{A}' \in {}^{\circ}\mathbb{C} \& F^{2\leftarrow}\mathfrak{A}' = \mathfrak{A}\}.$$

First we show that G^3 is precombinatorial, then that G^3 induces $F^1 \circ F^2$.

Suppose $G^3(\mathfrak{A}) \cap G^3(\mathfrak{B}) \neq \emptyset$. By requirement 1) of Definition 7.1 the values of G^1 are disjoint so $G^1(\mathfrak{A}') \subseteq G^3(\mathfrak{A}) \cap G^3(\mathfrak{B})$ for some $\mathfrak{A}' \in {}^{\circ}\mathbb{C}'$. But then $F^{2\leftarrow}\mathfrak{A}' = \mathfrak{A}$ and $F^{2\leftarrow}\mathfrak{A}' = \mathfrak{B}$. So $\mathfrak{A} = \mathfrak{B}$. Now Definition 7.1, 3) for G^3 follows from Definition 7.1, 3) for G^1. It remains to be shown that Definition 7.1, 2) holds of G^3. We must show that if $p: \mathfrak{A} \to \mathfrak{B}$ is an isomorphism of ${}^{\circ}\mathbb{C}$, then $G^3(\mathfrak{A})$ and $G^3(\mathfrak{B})$ are equinumerous. Compare $\bigcup \{G^1(\mathfrak{A}'): \mathfrak{A}' \in {}^{\circ}\mathbb{C}' \& F^{2\leftarrow}\mathfrak{A}' = \mathfrak{A}\}$ with $\bigcup \{G^1(\mathfrak{B}'): \mathfrak{B}' \in {}^{\circ}\mathbb{C}' \& F^{2\leftarrow}\mathfrak{B}' = \mathfrak{B}\}$. Observe that by Definition 7.1, 1) for G^1 these are disjoint unions. By Definition 7.1, 3) for G^1, if $\mathfrak{A}'$ is isomorphic to $\mathfrak{B}'$ then $G^1(\mathfrak{A}')$ is equinumerous with $G^1(\mathfrak{B}')$. Now by Lemma 7.8 restrictions of Fp match $\{\mathfrak{A}' \in {}^{\circ}\mathbb{C}': F^{2\leftarrow}\mathfrak{A}' = \mathfrak{A}\}$ with $\{\mathfrak{B}' \in {}^{\circ}\mathbb{C}: F^{2\leftarrow}\mathfrak{B}' = \mathfrak{B}\}$. Thus G^3 is precombinatorial. The combinatorial functor F^3 induced by G^3 is

$$F^3(\mathfrak{A}) = \mathrm{cl} \bigcup \{G^3(\mathfrak{A}'): \mathfrak{A}' \in {}^{\circ}\mathbb{C} \& \mathfrak{A}' \subseteq \mathfrak{A}\}$$

$$= \mathrm{cl} \bigcup \{G^1(\mathfrak{A}'): \mathfrak{A}' \in {}^{\circ}\mathbb{C}' \& F^{2\leftarrow}\mathfrak{A}' \subseteq \mathfrak{A}\}.$$

We compare this with the definition of F^1 from G^1.

$$F^1\big(F^2(\mathfrak{A})\big)=\mathrm{cl}\,\bigcup\,\{G^1(\mathfrak{A}'):\ \mathfrak{A}'\in{}^\circ\mathbb{C}'\ \text{and}\ \mathfrak{A}'\subseteq F^2(\mathfrak{A})\}\,.$$

Note that $F^{2\leftarrow}\,\mathfrak{A}'\subseteq\mathfrak{A}$ if and only if $\mathfrak{A}'\subseteq F^2(\mathfrak{A})$. This shows F^3 is $F^1\circ F^2$, and proves $F^1\circ F^2$ is a strict combinatorial functor. $\square$

Lemma 7.10 (Factorization Lemma). *Let $\mathfrak{M}'$ have dimension and degree 1. Let $\mathbb{C}=\mathbb{C}(\mathfrak{M})$, $\mathbb{C}'=\mathbb{C}(\mathfrak{M}')$. Let $F\colon\mathbb{C}\to\mathbb{C}'$ be a strict combinatorial functor. Suppose that $f'\colon E\to\mathfrak{M}'$ is an injection with range a basis for $\mathfrak{M}'$, let $V'\colon\mathbb{S}\to\mathbb{C}'$ be the closure functor (Corollary 7.5). Then there is a (strict) combinatorial functor $H\colon\mathbb{C}\to\mathbb{S}$ such that the restriction to the categories of inclusion maps of $F\colon\mathbb{C}\to\mathbb{C}'$ and $V'\circ H\colon\mathbb{C}\to\mathbb{C}'$ are naturally equivalent.*

Proof. Let $G\colon{}^\circ\mathbb{C}\to\mathscr{P}(\mathfrak{M}')$ be the precombinatorial operator inducing F. Let $p\colon\bigcup\{G(\mathfrak{A}'):\mathfrak{A}'\in{}^\circ\mathbb{C}\}\to f'(E)$ be an injection. Let $G'\colon{}^\circ\mathbb{C}\to\mathscr{P}(E)$ be the precombinatorial operator such that for $\mathfrak{A}\in{}^\circ\mathbb{C}$,

$$G'(\mathfrak{A})=(f')^{-1}\,p\,G(\mathfrak{A})\,.$$

Let $H\colon\mathbb{C}\to\mathbb{S}$ be the strict combinatorial functor induced by G'. Note $pG\colon{}^\circ\mathbb{C}\to\mathscr{P}(E)$ is a precombinatorial operator which induces the same map on isomorphism types objects in ${}^\circ\mathbb{C}$ to E^+ as does G.

Now for any $\mathfrak{A}\in{}^\circ\mathbb{C}$,

$$\begin{aligned}
(V'\circ H)(\mathfrak{A})&=\mathrm{cl}\big(f'\,H(\mathfrak{A})\big)\\
&=\mathrm{cl}\big(\bigcup\{f'\,G'(\mathfrak{A}'):\ \mathfrak{A}'\subseteq\mathfrak{A}\,\&\,\mathfrak{A}'\in{}^\circ\mathbb{C}\}\big)\\
&=\mathrm{cl}\big(\bigcup\{pG(\mathfrak{A}):\ \mathfrak{A}'\subseteq\mathfrak{A}\,\&\,\mathfrak{A}'\in{}^\circ\mathbb{C}\}\big)\,.
\end{aligned}$$

So pG induces $V'\circ H$. Since G, pG induce the same function on finite isomorphism types, and respectively induce F and $V'\circ H$, by Theorem 7.6, F and $V'\circ H$ are naturally equivalent on inclusion maps. (In fact, for all $\mathfrak{A}$, $(V'\circ H)(\mathfrak{A})=(\mathrm{cl}\,p)(F(\mathfrak{A}))$.) $\square$

Theorem 7.11. *Let $\mathfrak{M}^1$, $\mathfrak{M}^2$ have dimension and degree 1. Let $\mathbb{C}^i=\mathbb{C}(\mathfrak{M}^i)$. Let $f^i\colon E\to\mathfrak{M}^i$ be an injection with range a basis for $\mathfrak{M}^i$. Let $V^i\colon\mathbb{S}\to\mathbb{C}^i$ be the corresponding closure functor. Then there exists a (strict) combinatorial functor $H\colon\mathbb{S}\to\mathbb{S}$ such that the restrictions to inclusion maps of $F\circ V^1\colon\mathbb{S}\to\mathbb{C}^2$ and $V^2\circ H\colon\mathbb{S}\to\mathbb{C}^2$ are naturally equivalent.*

Proof. $F\circ V^1\colon\mathbb{S}\to\mathbb{C}^2$ is a strict combinatorial functor by Theorem 7.9. By Lemma 7.10 there is an $H\colon\mathbb{S}\to\mathbb{S}$ with $F\circ V^1$ and $V^2\circ H$ naturally equivalent. $\square$

Example 7.11. Direct sum of vector spaces can be constructed as a combinatorial functor, but not as a strict combinatorial functor. Let U be a basis for the vector space $\mathfrak{M}$, let $\mathbf{V}$ be the category arising from $\mathfrak{M}$. Let $F^1\colon U \to U$, $F^2\colon U \to U$ be injections such that U is the disjoint union of $F^1(U)$ and $F^2(U)$. Let $L^i\colon \mathfrak{M} \to \mathfrak{M}$ be the extension of F^i to a linear transformation. For objects $\mathfrak{A}$, $\mathfrak{B}$ in $\mathbf{V}$, define $F(\mathfrak{A}, \mathfrak{B})$ as the subspace spanned by $L^1\,\mathfrak{A}$ and $L^2\,\mathfrak{B}$. Define $F(p,q)$ for morphisms p,q so that $F(p,q)(L^1\,a + L^2\,b) = L^1\,p\,a + L^2\,q\,b$ for $a \in \mathfrak{A}$, $b \in \mathfrak{B}$. A trivial exercise shows this is a combinatorial functor. F is not strict. If it were, let G be its precombinatorial operator. Let b_1, b_2 be independent, let $\mathfrak{A}^1$, $\mathfrak{A}^2$, $\mathfrak{A}^3$ be the one dimensional subspaces spanned by $b_1, b_2, b_1 + b_2$ respectively. Let $\mathfrak{O}$ be the zero dimensional subspace. Then $F(\mathfrak{O}, \mathfrak{O}) = \mathfrak{O}$ implies $G(\mathfrak{O}, \mathfrak{O})$ is empty. So $F(\mathfrak{A}^i, \mathfrak{O})$ is one dimensional implies $G(\mathfrak{A}^i, \mathfrak{O})$ has cardinality 1. But $F(\mathfrak{A}^3, \mathfrak{O})$ is spanned by $F(\mathfrak{A}^2, \mathfrak{O})$ and $F(\mathfrak{A}^1, \mathfrak{O})$, so $G(\mathfrak{A}^3, \mathfrak{O})$ must be empty to be independent of $G(\mathfrak{A}^2, \mathfrak{O})$, $G(\mathfrak{A}^1, \mathfrak{O})$ which is a contradiction.

8. Strict Combinatorial Functions

We develop an arithmetic tool appropriate for the analysis of strict combinatorial functors between categories of the form $\mathbb{C}(\mathfrak{M})$ at least when $\mathfrak{M}$ is a countably infinite atomic model in which finitely generated algebraically closed sets are finite and $\mathfrak{M}$ has dimension and degree 1. The most important examples are $\mathbb{S}$, and $\mathbb{V}$ over a finite field. The above properties are assumed for all categories of this section. Let $\mathbb{C} = \mathbb{C}(\mathfrak{M})$, $\mathbb{C}' = \mathbb{C}(\mathfrak{M}')$.

Definition 8.1. A functor $F\colon \mathbb{C} \to \mathbb{C}'$ is said to be *finitary* if F maps $°\mathbb{C}$ to $°\mathbb{C}'$. Each finitary combinatorial functor $F\colon \mathbb{C} \to \mathbb{C}'$ gives rise to a function $F^{\#}\colon E \to E$.

$$F^{\#}(\dim \mathfrak{A}) = \dim F(\mathfrak{A}) \qquad \text{for all } \mathfrak{A} \in °\mathbb{C}.$$

A function $f\colon E \to E$ arising as $F^{\#}$ from a finitary combinatorial functor $F\colon \mathbb{C} \to \mathbb{C}'$ is said to be $\mathbb{C}$ to $\mathbb{C}'$ *combinatorial*; or if $\mathbb{C} = \mathbb{C}'$, simply $\mathbb{C}$-*combinatorial*. A function $f\colon E \to E$ arising as $F^{\#}$ from a finitary strict combinatorial $F\colon \mathbb{C} \to \mathbb{C}'$ is said to be a *strict $\mathbb{C}$-combinatorial function*. (It will shortly be shown that the specific $\mathbb{C}'$ is irrelevant.) The $\mathbb{S}$-combinatorial functions and the strict $\mathbb{S}$-combinatorial functions coincide and were introduced by Myhill [1958].

Lemma 8.2. *Composition of functors induces composition of combinatorial functions; in particular $\mathbb{C}$-combinatorial functions are closed under composition. Strict $\mathbb{C}$-combinatorial functions are closed under composi-*

tion. (*More generally, if* $f(v_1, \ldots, v_n)$ *is a strict* $\mathbb{C}$-*combinatorial function and* $g_1(u_1, \ldots, u_k), \ldots, g_n(u_1, \ldots, u_k)$ *are* $\mathbb{C}$-*combinatorial, then* $f(g_1, \ldots, g_n)$ *is a strict* $\mathbb{C}$-*combinatorial function.*)

Proof. We confine ourselves to one variable. If $F: \mathbb{C} \to \mathbb{C}'$, $G: \mathbb{C}' \to \mathbb{C}''$ are combinatorial, so is $G \circ F: \mathbb{C} \to \mathbb{C}''$ by Theorem 3.2. If G is in addition strict, then so is F by Theorem 7.9. Finally

$$(G \circ F)^{\#}(\dim \mathfrak{A}) = \dim(G \circ F(\mathfrak{A})) = G^{\#}(\dim F(\mathfrak{A})) = G^{\#}(F^{\#}(\mathfrak{A}))$$

for $\mathfrak{A} \in {}^{\circ}\mathbb{C}$. $\square$

In $\mathbb{C} = \mathbb{C}(\mathfrak{M})$ let $\begin{bmatrix} n \\ i \end{bmatrix}$ be the number of i dimensional algebraically closed subsets of an n dimensional algebraically closed set. (We use a notation like $\begin{bmatrix} n \\ i \end{bmatrix}_{\mathbb{C}}$ if dependence on $\mathbb{C}$ needs to be shown.)

Theorem 8.3 (Myhill Normal Form). *A function* $f: E \to E$ *is strict* $\mathbb{C}$-*combinatorial if, and only if, there is a function* $c: E \to E$ *such that for all n in E*

$$f(n) = \sum_{i=0}^{n} c(i) \begin{bmatrix} n \\ i \end{bmatrix}.$$

Proof. Suppose f is given in the form $\sum c(i) \begin{bmatrix} n \\ i \end{bmatrix}$, where all $c(i) \in E$. Suppose $\mathbb{C} = \mathbb{C}(\mathfrak{M}')$ is given. By Lemma 7.2 we produce a (Myhill) precombinatorial operator $G: {}^{\circ}\mathbb{C} \to \mathscr{P}(\mathfrak{M}')$ such that whenever $\mathfrak{A} \in {}^{\circ}\mathbb{C}$ has dimension i, $G(\mathfrak{A})$ has $c(i)$ elements. Let F be the strict combinatorial functor $F: \mathbb{C} \to \mathbb{C}'$ induced by G as in Theorem 7.4. Note that F is finitary since $c(i) \in E$ and $\begin{bmatrix} n \\ i \end{bmatrix}$ is finite for finite n and i. All parts of Theorem 8.3 will be verified if we show that for any strict finitary combinatorial functor F induced by a precombinatorial operator G, if $c(\dim \mathfrak{A}) = \dim G(\mathfrak{A})$ for all $\mathfrak{A} \in {}^{\circ}\mathbb{C}$, then

$$F^{\#}(n) = \sum_{i=0}^{n} c(i) \begin{bmatrix} n \\ i \end{bmatrix}.$$

But if $\mathfrak{A}$ has dimension n, remember that

$$F(\mathfrak{A}) = \operatorname{cl} \bigcup \{G(\mathfrak{A}'): \mathfrak{A}' \subseteq \mathfrak{A}, \mathfrak{A}' \in {}^{\circ}\mathbb{C}\}.$$

So since $\bigcup \{G(\mathfrak{A}'): \mathfrak{A}' \subseteq \mathfrak{A}, \mathfrak{A}' \in {}^{\circ}\mathbb{C}\}$ is independent (Definition 7.1, Part 3) and is a pairwise disjoint union (Definition 7.1, Part 2) we get

$$F^{\#}(n) = \sum \{\operatorname{card} G(\mathfrak{A}'): \mathfrak{A}' \subseteq \mathfrak{A}, \mathfrak{A}' \in {}^{\circ}\mathbb{C}\}.$$

But $\dim \mathfrak{A}' = \dim \mathfrak{A}'' = i$ implies that $\mathfrak{A}'$ is isomorphic to $\mathfrak{A}''$ by an isomorphism of $\mathbb{C}$, so (by Definition 7.1, Part 2) $G(\mathfrak{A}')$ and $G(\mathfrak{A}'')$ have the same cardinality. Therefore all i dimensional algebraically closed subsets $\mathfrak{A}'$ of $\mathfrak{A}$ make the same contribution $c(i) = \operatorname{card} G(\mathfrak{A}')$ to the sum. This proves the formula. □

We call c the *Stirling coefficient function* for f and $\sum_{i=0}^{\infty} c(i) \begin{bmatrix} n \\ i \end{bmatrix}$ the *combinatorial series* for f.

An easy induction demonstrates that for a given $\mathbb{C}$, a function $f \colon E \to E$ has at most one expansion $f(n) = \sum_{i=0}^{n} c(i) \begin{bmatrix} n \\ i \end{bmatrix}_{\mathbb{C}}$ (possibly none).

Example 8.1. In $\mathbb{S}$, $\begin{bmatrix} n \\ i \end{bmatrix}$ is simply the combinatorial coefficient $\binom{n}{i} = \dfrac{n!}{i!\,(n-i)!}$. In this case any $f \colon E \to E^* = \{0, \pm 1, \pm 2, \ldots\}$ has a unique expansion $f(n) = \sum_{i=0}^{n} c(i) \binom{n}{i}$ when $c \colon E \to E^*$. Here $c(n) = (\Delta^n f)(n)$, where $(\Delta f)(n) = f(n+1) - f(n)$.

Example 8.2. If $\mathbb{V}$ is the vector space category over a finite field, $\begin{bmatrix} n \\ i \end{bmatrix}$ is the number of i dimensional subspaces of an n dimensional space. See Burnside, Theory of Groups [1883], p. 109–111, for a calculation.

The generalization of these ideas to several variables is straightforward. If $F \colon \mathbb{C}_0 \times \cdots \times \mathbb{C}_k \to \mathbb{C}$ is a strict combinatorial functor, then

$$F^{\#}(n_0, \ldots, n_k) = \sum_{i_0 = 0}^{n_0} \cdots \sum_{i_k = 0}^{n_k} c(i_0, \ldots, i_k) \begin{bmatrix} n_0 \\ i_0 \end{bmatrix}_{\mathbb{C}_0} \cdots \begin{bmatrix} n_k \\ i_k \end{bmatrix}_{\mathbb{C}_k},$$

where $\mathbb{C} \colon \times^k E \to E$ and the $\begin{bmatrix} n_j \\ i_j \end{bmatrix}$ are multiplied together.

Usually we consider only the case all $\mathbb{C}_i$ are $\mathbb{C}$.

Example 8.3. $x + y = \binom{x}{1} + \binom{y}{1}$, $x \cdot y = \binom{x}{1} \cdot \binom{y}{1}$ are $\mathbb{S}$-combinatorial. If $c(i)$ is the number of permutations of an i element set moving every element, then $x! = \sum c(i) \binom{x}{i}$, so $x!$ is $\mathbb{S}$-combinatorial. If $c(i, j)$ is the number of functions from a j element set onto an i element set, then $(x + 1)^y = \sum c(i, j) \binom{x}{i} \binom{y}{j}$, so $(x + 1)^y$ is $\mathbb{S}$-combinatorial.

Example 8.4. A strict $\mathbb{V}$-combinatorial function $f(x)$ which is not constant must increase at least as fast as $\begin{bmatrix} x \\ 1 \end{bmatrix}_{\mathbb{V}}$. So $x, x + y$ are *not* strictly $\mathbb{V}$-combinatorial functions. But they are $\mathbb{V}$-combinatorial, being induced respectively by the identity functor and the direct sum functor (Example 7.1).

Example 8.5. If $V: \mathbb{S} \to \mathbb{C}$ is the closure functor of Corollary 7.5, then $V^{\#}$ is the identity function on E.

For dim $V(\mathfrak{A}) =$ dim $\mathfrak{A} =$ card $\mathfrak{A}$.

Theorem 8.4. *Every strict combinatorial function is* $\mathbb{S}$-*combinatorial.*

Proof. Let $\mathbb{C} = \mathbb{C}^1$, let f be strict $\mathbb{C}^1$ to $\mathbb{C}^2$ combinatorial and let $F: \mathbb{C}^1 \to \mathbb{C}^2$ be a strict combinatorial functor inducing f so that $f = F^{\#}$. By Theorem 7.11 there is a (strict) combinatorial functor $H: \mathbb{S} \to \mathbb{S}$ such that $F \circ V^1$ and $V^2 \circ H$ are naturally equivalent on the subcategories of inclusion maps. Then obviously for $\mathfrak{A} \in {}^{\circ}\mathbb{C}^1$, $F\big(V^1(\mathfrak{A})\big)$ is isomorphic to $V^2\big(H(\mathfrak{A})\big)$. Taking dimensions of each side, if $\mathfrak{A}$ has dimension n,

$$F^{\#}\big(V^{1\#}(n)\big) = V^{2\#}\big(H^{\#}(n)\big).$$

By Examples 8.5, $F^{\#} = H^{\#}$ si $f = F^{\#}$ is $\mathbb{S}$-combinatorial. $\square$

Corollary 8.5. $\begin{bmatrix} n \\ i \end{bmatrix}_{\mathbb{C}}$ *is* $\mathbb{S}$-*combinatorial.* $\square$

Now consider strict combinatorial functors $F: \mathbb{C} \to \mathbb{C}'$ where F is not necessarily finitary. In a formal sense the proof of Theorem 8.3 still works, where instead of a function on integers, $F^{\#}$, we have instead a "formal combinatorial series"

$$\sum c(i) \begin{bmatrix} n \\ i \end{bmatrix}, \quad \text{where} \quad c: E \to E^+,$$

and
$$c(\text{dim } \mathfrak{A}) = \text{card } G(\mathfrak{A}), \quad \text{for } \mathfrak{A} \in {}^{\circ}\mathbb{C},$$

where G is a precombinatorial operator inducing F. Note this definitely allows $c(n) = \aleph_0$. Then there is a rule for composing series corresponding to composition of strict functors and generalizing composition of functions. In the case of $\mathbb{S}$ Nerode [1966a] wrote out the calculation. Of course the algebra of strict combinatorial functions is naturally imbedded preserving composition in this algebra of formal combinatorial series. Finally, each combinatorial series can be regarded as an $\mathbb{S}$-combinatorial series generalizing Theorem 8.4.

The arithmetic of $\mathbb{S}$-combinatorial series is used to investigate non-Dedekind types. Since we mostly investigate only Dedekind types, further development of combinatorial series is omitted.

If $\mathfrak{A}$ is an object of $\mathbb{C}$ then the classical isomorphism type $\langle \mathfrak{A} \rangle$ of $\mathfrak{A}$ consists of all codomains of isomorphisms of $\mathbb{C}$ with domain $\mathfrak{A}$. If $F: \mathbb{C}_0 \times \cdots \times \mathbb{C}_k \to \mathbb{C}$ is a (combinatorial) functor then F induces a map F on classical isomorphism types given by

$$\mathsf{F}(\langle \mathfrak{A}_0 \rangle, \ldots, \langle \mathfrak{A}_k \rangle) = \langle F(\mathfrak{A}_0, \ldots, \mathfrak{A}_k) \rangle$$

for objects $\mathfrak{A}_0 \in \mathbb{C}_0, \ldots, \mathfrak{A}_k \in \mathbb{C}_k$. If, in addition, F is a strict, finitary combinatorial functor then F induces a function $F^\#$ on the natural numbers according to Definition 8.1.

Corollary 8.6. *Suppose* $F, G, \ldots$ *are strict finitary combinatorial functors. Then any identity in* E *between* $F^\#, G^\#, \ldots$ *yields a corresponding identity in classical isomorphism types between* $\mathsf{F}, \mathsf{G}, \ldots$ *and conversely.*

Proof. The transformations $F \leadsto \mathsf{F}$ and $F \leadsto F^\#$ both commute with composition so it suffices to show that if F, G are strict finitary combinatorial functors then $\mathsf{F} = \mathsf{G}$ if, and only if, $F^\# = G^\#$. But if $\mathsf{F} = \mathsf{G}$ then for $\mathfrak{A}$ of finite dimension n, $\mathsf{F}(\langle \mathfrak{A} \rangle) = \mathsf{G}(\langle \mathfrak{A} \rangle)$ so $\dim \mathsf{F}(\langle \mathfrak{A} \rangle) = \dim \mathsf{G}(\langle \mathfrak{A} \rangle)$ so $F^\#(n) = G^\#(n)$. Conversely if $F^\# = G^\#$ then by Theorem 7.6, F and G are naturally equivalent on objects so $\mathsf{F} = \mathsf{G}$. $\square$

We note that the above corollary also holds when $F^\#, G^\#, \ldots$ are strict combinatorial series as well as when they are strict combinatorial functions.

Part IV. Recursive Equivalence

9. Suitable Categories

Let L be a complete effectively presented countable first order language with identity. Let $L(C)$ be the extension of L obtained by adding a fully effective countably infinite set of individual constants. If T' is a complete theory in $L(C)$ and T a theory in L then T' is said to be a *witness completion* of T if $T' \supseteq T$ and whenever $\exists v_n \, \phi(v_n)$ is in T, then $\phi(c)$ is in T' for some $c \in C$. The Henkin-Hasenjaeger proof of completeness of first order logic has an obvious effective version obtained by simply making all listings effective (see Mendelson [1964] exercise p. 65 and Proposition 2.12). There is also an obvious relativized version.

Theorem 9.1 (Effective Completeness Theorem). *Let T be a recursively enumerable theory in L then T is decidable if, and only if, there is a recursively enumerable sequence of recursive witness completions $T_0, T_1, \ldots$ of T such that for all ϕ, ϕ is a consequence of T if, and only if, for all i, ϕ is a consequence of T_i.* $\square$

The significance of witness completions classically is that they give an alternate syntactic version of the semantic notion of model which is obtained by introducing a name in C for every element of the model. Likewise recursive witness completions give an alternate syntactic version of the semantic notion of recursively presented model.

Definition 9.2. A model $\mathfrak{M}$ of a theory T is said to be *recursively presented* if the universe of $\mathfrak{M}$ is an initial segment of the natural numbers and the satisfaction relation "a satisfies ϕ in $\mathfrak{M}$" is recursive (where a is a finite sequence of elements of $|\mathfrak{M}|$ and ϕ a formula of L).

Corollary 9.3. *Every decidable theory T has a recursively presented model.* $\square$

This notion of recursively presented model is stronger than some in the literature. What is required is recursiveness for *all* first order formulae, not just for atomic formulae. (If we adopt the Morley [1965] stratagem of having an atomic formula equivalent to each formula this distinction is blurred.) However, simply having a recursive presentation does not mean that all simple questions can be answered.

Definition 9.4. A theory T is said to have *decidable atoms* if for any n and any formula $\phi(v_0, \ldots, v_{n-1})$ (with at most $v_0, \ldots, v_{n-1}$ free) there is a uniform effective procedure which decides whether ϕ is an atom of $B_n(T)$ or not.

Vaught's Theorem 4.2 immediately yields an effective version.

Theorem 9.5. *Let T be a complete theory. Suppose $B_n(T)$ is atomistic for all n, T is decidable and has decidable atoms. Then T has a recursively presented atomic model and any two recursively presented atomic models of T are isomorphic by a recursive isomorphism.* $\square$

(Here a recursive isomorphism means an isomorphism which is a recursive function.)

Now suppose T is complete and decidable, $B_n(T)$ is atomistic for all n, and T has decidable atoms and an infinite model. If $\mathfrak{M}$ is the recursively presented atomic model of T, then the operation of taking the algebraic closure in $\mathfrak{M}$ of a set A is effective in the sense that, from an enumeration of A we can uniformly obtain an enumeration of $\mathrm{cl}(A)$. For if $a_0, \ldots, a_{n-1} \in A$ are given, then we first find the atom $\phi(v_0, \ldots, v_{n-1})$ they satisfy, then effectively list all atoms $\psi(v_0, \ldots, v_n)$ and natural numbers k such that

$$T \vdash \phi(v_0, \ldots, v_{n-1}) \to \exists !^k v_n \, \psi(v_0, \ldots, v_n)$$

and for each of these list all a_n such that $a_0, \ldots, a_n$ satisfy ψ. However, if A is finite this does not give an explicit index of $\mathrm{cl}(A)$. If T is $\aleph_0$-categorical then each $B_n(T)$ is finite and the situation is somewhat improved. For each ϕ we can list the formulae ψ and, since we only need to look at atoms, we can just list the ψ which are atoms and then check to see when the disjunction of the atoms ψ so far enumerated is a consequence of T (that is, is the unit of $B_{n+1}(T)$). When the disjunction has this property all the atoms have been enumerated. However, this does not give us a bound on the natural number k. So, since we need algebraic closure to be effective on explicit indices of finite sets we introduce one last assumption.

Theorem 9.6. *Suppose T is complete, decidable, and $\aleph_0$-categorical and has decidable atoms. Then the map $A \leadsto \mathrm{cl}(A)$ is effective on explicit indices (in the recursively presented model) if, and only if, for any atoms $\phi(v_0, \ldots, v_{n-1})$, $\psi(v_0, \ldots, v_n)$ in $B_{n+1}(T)$ such that $T \vdash \phi \to \exists v_n \psi$ we can effectively decide whether or not there is a $k > 0$ such that $T \vdash \phi \to \exists !^k v_n \psi$.*

Proof. Suppose $A \leadsto \mathrm{cl}\, A$ is effective on explicit indices and ϕ, ψ are given atoms. Find $a_0, \ldots, a_{n-1}$ in $\mathfrak{M}$ satisfying ϕ and compute $\mathrm{cl}\{a_0, \ldots, a_{n-1}\}$. For each of its elements a check whether $a_0, \ldots, a_{n-1}$, a satisfies ψ. If some element does there is a $k > 0$ such that $T \vdash \phi \to \exists !^k v_n \psi$. If not, there is no such k.

Conversely, if we can effectively decide whether or not there is a $k>0$ then, by the remarks preceding this theorem, we can give an explicit index of cl A from one for A, since we have finitely many atoms ψ which we have to test. This completes the proof. $\square$

We now note that if $\mathfrak{M}$ has dimension then the computations with bases are effective.

At last we are ready for the definition of suitable categories $\mathbb{C}(\mathfrak{M})$ used throughout the rest of the book.

Let $\mathfrak{M}$ be a recursively presented model. Then the objects of $\mathbb{C}(\mathfrak{M})$ are the algebraically closed subsets of $\mathfrak{M}$. The morphisms of $\mathbb{C}(\mathfrak{M})$ are those elementary monomorphisms p such that p has a one-one partial recursive extension.

$^{\circ}\mathbb{C}(\mathfrak{M})$ is the full sub-category of finite morphisms and finite objects of $\mathbb{C}(\mathfrak{M})$.

Definition 9.7. A category $\mathbb{C}$ is said to be *suitable* if $\mathbb{C}=\mathbb{C}(\mathfrak{M})$ for some recursively presented, infinite, $\aleph_0$-categorical model $\mathfrak{M}$ of a complete theory T that is decidable, has decidable atoms and is such that algebraic closure is effective on explicit indices of finite sets.

It is worth noting that, with suitable effectivity conditions, a complete theory T such that each $B_n(T)$ is atomistic and in which finitely generated algebraically closed sets are finite will work for most of the results of the monograph.

Notational convention. From now on $\mathbb{S}$, $\mathbb{L}$, $\mathbb{V}$, $\mathbb{B}$ are to be used for the categories arising from the $\aleph_0$-categorical theories of infinite sets with equality, dense unbordered linear orderings, infinite dimensional vector spaces over a fixed finite field and atomless Boolean algebras, respectively.

10. Bridge

The original recursive equivalence types of Dekker [1955] were introduced as analogues of cardinal numbers in set theory. The central reason for their study is a desire to understand the non-standard Dedekind notion of finiteness in all its manifestations. Beside that there was the objective of developing a theory of invariants for recursion theory like that of cardinals for set theory. In set theory cardinals make crude but useful distinctions. (Thus one may distinguish sets in analysis by showing they have different cardinalities.) In the original work arbitrary sets were replaced by sets of natural numbers, arbitrary one-one functions by one-one partial recursive functions and equivalence classes of sets under one-one functions (cardinals, that is) by recursive equivalence types of natural numbers (RETs). Sets which cannot be mapped one-one onto a proper subset of themselves (that is, sets which are not infinite according to Dedekind's definition) are then replaced by sets of natural numbers which cannot be mapped one-one onto a proper subset by a one-one partial recursive function (these are effectively Dedekind finite sets or isolated sets). Then Dedekind cardinals have as analogues Dedekind RETs (otherwise called isols).

The analogue of cardinal arithmetic proved quite fruitful and the methods introduced (in particular Myhill's combinatorial functions and Nerode's frames) themselves contributed to the theory of cardinals in set theory without the full axiom of choice (Ellentuck [1966]). This was particularly so after the appearance of Cohen's method of forcing. Further, the methods when used in conjunction with the priority method suffice to obtain a deep theory of recursive equivalence types of cosimple sets (Hay [1966]). This showed that the methods are appropriate for a context where both the one-one functions and the sets are taken from equally narrow and classically equinumerous collections.

Since one-one functions preserve so little structure and structure preserving maps are the heart of algebra, it was enticing to envisage a corresponding theory of structure preserving maps and their corresponding recursive isomorphism types. The first extensive attempt was made on linear orderings (Crossley [1969]), the basic idea having been independently discovered by Manaster (unpublished). Hence linear orderings of the classical theory of order types were replaced by suborderings of the rationals and order preserving functions were replaced by one-one partial recursive order preserving functions. Classical order types were then replaced by "constructive" order types.

In Crossley's work two features became apparent. The first is that it is a great advantage to consider only substructures of a fixed, effectively given structure (here suborderings of the rationals) and their types, rather than structures with not necessarily recursive atomic relations which might not be extendable to effective structures. This is because recursion theory of structures proceeds by uniform effective methods on them, and without effective relations there are practically no effective uniformities present.

The second feature is that the notion of Dedekind finiteness should remain the same, that is to say, a subordering is Dedekind if its *domain* is Dedekind. (In recursion theory this means a subordering of the rationals is Dedekind if its domain contains no infinite r.e. set; in set theory a structure is Dedekind if its domain has a Dedekind cardinal.) The *a priori* reason for this is to study structures of Dedekind cardinality to see how they behave. The *a posteriori* reason is that this is the only evident way to permit deduction of properties of Dedekind structures from properties of finite structures. In this way we can think of Dedekind structures more or less directly as "limits" of finite structures and Dedekind finiteness as a "non-standard" version of ordinary finiteness.

A substantial exploratory investigation into types of vector spaces over an infinite recursive field is due to Dekker [1969]. (There the definition of Dedekind vector space is altered to allow for the presence of infinite recursively enumerable one dimensional subspaces. The present monograph encompasses in this regard only the case of vector spaces over a finite field where this situation does not arise.) But from this Dekker gave us an important insight, namely, that recursive equivalence of structures should be based on isomorphisms extendable to one-one partial recursive functions. (This is more general than one-one partial recursive functions p which are isomorphisms from domain p onto range p. This also had the consequence that recursive equivalence for linear orderings as treated in this monograph is not quite identical with the corresponding concept in Crossley [1969] although few theorems there are affected.) The notion of recursive equivalence is appropriate because a fundamental device for finding isomorphisms between Dedekind structures is the piecing together of infinitely many effectively given finite isomorphisms into one large one-one partial recursive function. But it is often the case that the isomorphisms put together are only compatible as one-one maps, not as structure preserving isomorphisms.

It was Peter Aczel who in conversations in 1965–1966 and in his regrettably unpublished dissertation [1966] first gave an account of combinatorial operators from the point of view of continuous homomorphisms. This was the first hint in print that a functorial point of view is profitable in RETs.

Finally, we observed quite early that with Dekker's theory of RETs, Crossley's of constructive order types and Dekker's of RETs of vector spaces over a finite field there is associated a fully effective, $\aleph_0$-categorical theory and infinite model. For RETs the theory is that of an infinite set with equality and the model the natural numbers with equality. For constructive order types the theory is that of a dense unbordered linear ordering and the model the rationals. With vector spaces over a fixed finite field the model is that of an $\aleph_0$-dimensional vector space over the field and its theory is the $\aleph_0$-categorical theory of that model. In this case however once a structure preserving map is given on a subset it has a unique extension to the vector subspace generated by that set.

This led us to consider the notion of the algebraic closure of a set in a model. For sets and linear orderings every subset is algebraically closed, for vector spaces the vector subspaces are the algebraically closed sets.

With this in mind it seemed natural to ask about the recursive isomorphism types of algebraically closed subsystems of a fully effective, but otherwise arbitrary, $\aleph_0$-categorical model. To this question the present monograph is devoted.

The fundamental attributes of algebraically closed subsets of a fully effective, $\aleph_0$-categorical model which make the theory work out are roughly as follows.

The assumption that the model is "fully effective" guarantees that uniform effective methods of recursion theory are applicable. $\aleph_0$-categorical models are, of course, atomic and the "fully effective" assumption means that computations with atoms (and with bases when we have a dimension as for vector spaces) such as occur in the preceding sections attain an effective character. The assumption of $\aleph_0$-categoricity ensures that finitely generated algebraically closed sets are finite and, together with the effectiveness assumption, guarantees that the computation of algebraic closure is fully effective. This uniformity is required when we approximate Dedekind structures by means of finite structures in a uniform manner. It is important that the models have sufficient room for internal duplication, in particular we have the Duplication Lemma which ensures that every infinite algebraically closed set has $2^{\aleph_0}$ classically isomorphic copies with a sort of effective presentation in the $\aleph_0$-categorical model. This is necessary for the production of counterexamples.

Finally $\aleph_0$-categorical models are universal for certain classes of countable structure. In the three examples we have mentioned these are very natural; for the classes we have considered are countable sets, linear orderings and vector spaces over a finite field. In the sequel we consider other such examples of natural classes, in particular, countable Boolean algebras.

(Perhaps surprisingly a perusal of the proofs appears to indicate that the principal property of $\aleph_0$-categorical models, namely that there are only finitely many n-types for each n, is never really necessary. The main assumption is that of a countable atomic model in which finitely generated algebraically closed sets are finite and all aspects of computing with algebraic closure are effective. However, for simplicity we restrict ourselves to the $\aleph_0$-categorical case throughout.)

11. Recursive Equivalence (Sets)

"The classical theory of cardinal numbers deals with a property of collections (their cardinality) which is preserved under all one-one mappings. The present study is concerned with a property of collections

(their effective equivalence type) which is preserved under all effective
one-one mappings. It may therefore be regarded as a constructive
analogue of Cantor's theory." (Introduction to Dekker and Myhill
[1960].)

Subsequently one of the present authors developed an analogue of
Cantor's theory of order types. In this monograph we go much further
and extend to the constructive analogues of isomorphism types of many
sorts of algebraic structure.

In the next three sections we review the cardinal and order type
analogues briefly. For fuller details the reader should see (in particular)
Dekker and Myhill [1960], Dekker [1966], Nerode [1959, 1966] for
the cardinal case and Crossley [1969] for the order case.

We consider the category $\mathbb{S}$ of sets of natural numbers. $\mathfrak{A}$ is said
to be *equivalent* to $\mathfrak{B}$ if there is a morphism in $\mathbb{S}$ (i.e. a one-one function)
mapping $\mathfrak{A}$ onto $\mathfrak{B}$. This relation is an equivalence relation and the
equivalence classes are the cardinals $0, 1, 2, \ldots, \aleph_0$.

Definition 11.1. $\mathfrak{A}$ is said to be *recursively equivalent* to $\mathfrak{B}$ if there
is a morphism in $\mathbb{S}$ which maps $\mathfrak{A}$ onto $\mathfrak{B}$ and which can be extended
to a one-one partial recursive function. We write $\mathfrak{A} \simeq \mathfrak{B}$ in this case.
Such a morphism is called a *recursive equivalence*.

Recursive equivalence is indeed an equivalence relation. (To see
that it is transitive, define $p \circ q(x) = y$ if, and only if, for some z, $q(x)$ is
defined and equals z and $p(z)$ is defined and equals y; then $p \circ q$ is a
one-one partial recursive function if p and q are.) We write $\langle \mathfrak{A} \rangle$ for
the equivalence class of $\mathfrak{A}$ under recursive equivalence. $\langle \mathfrak{A} \rangle$ is called
the recursive equivalence type (RET) of $\mathfrak{A}$.

Let $\Omega(\mathbb{S})$ denote the collection of all RETs. $\Omega(\mathbb{S})$ has the cardinality
c of the continuum since there are c sets and only $\aleph_0$ recursive equiv-
alences, but the finite RETs (RETs of finite sets) may be identified
with the finite cardinals.

Dedekind's definition of infinite is as follows: $\mathfrak{A}$ is infinite if there
is a one-one map of $\mathfrak{A}$ onto a proper subset of $\mathfrak{A}$. This motivates the
following definition.

Definition 11.2. $\mathfrak{A}$ is said to be *effectively Dedekind infinite* if there
is a one-one function p with a one-one partial recursive extension such
that p maps $\mathfrak{A}$ onto a proper subset of $\mathfrak{A}$.

$\mathfrak{A}$ is said to be *effectively Dedekind finite* if it is not effectively
Dedekind infinite.

Lemma 11.3. $\mathfrak{A}$ *is effectively Dedekind infinite if, and only if, $\mathfrak{A}$ contains an infinite recursively enumerable subset.*

Proof. If $p: \mathfrak{A} \simeq \mathfrak{B} \subset \mathfrak{A}$ take $x \in \mathfrak{A} - \mathfrak{B}$ then $\{p^n(x): n = 0, 1, 2, ...\}$ is an infinite recursively enumerable subset of $\mathfrak{A}$ (where $p^{n+1}(x) = p(p^n(x))$). Conversely, if $\mathfrak{A}$ contains an infinite recursively enumerable subset, it contains an infinite recursive subset $\mathfrak{B}$ such that $\mathfrak{B} = \{f(n): n \in E\}$ where f is a one-one recursive function. Set $p(x) = x$ if $x \notin \mathfrak{B}$ and $p(f(n)) = f(n+1)$ then $p: \mathfrak{A} \to \mathfrak{A} - \{f(0)\}$. $\square$

Definition 11.4. An RET A is said to be *Dedekind* if some (or equivalently every) $\mathfrak{A} \in A$ is effectively Dedekind finite. We write $\Lambda(\mathbb{S})$ for the collection of Dedekind $\mathbb{S}$-RETs.

Dekker originally called Dedekind RETs *isols*.

The theory of RETs has as a major concern the transfer of theorems from finite cardinals to Dedekind RETs. That infinite Dedekind RETs exist follows from the next lemma whose proof can readily be obtained from Corollary 12.3 below.

Lemma 11.5. *There exist c Dedekind RETs which are infinite.* $\square$

12. Recursive Equivalence (Linear Orderings)

Cardinal numbers have an effective analogue in the recursive equivalence types of Dekker. Order types have an effective analogue too. The first author developed the theory of properties of (countable) linear orderings under one-one partial recursive order isomorphisms (Crossley [1963], [1965], [1969]), though we note that the original notion is also independently due to Manaster (unpublished).

We take as $\mathbb{L}$ the category of all subsets (with the induced order) of $\mathfrak{Q}$, the rational numbers with the natural order $\leq$ (under suitable coding), since any (countable) dense unbordered linear order is isomorphic to $\mathfrak{Q}$.

Definition 12.1. If $\mathfrak{A}, \mathfrak{B} \in \mathbb{L}$ then $\mathfrak{A}$ is said to be *recursively $\mathbb{L}$-equivalent* to $\mathfrak{B}$ if there is a morphism in $\mathbb{L}$ which maps $\mathfrak{A}$ onto $\mathfrak{B}$ and which can be extended to a one-one partial recursive function. We write $\mathfrak{A} \simeq \mathfrak{B}$ in this case or $p: \mathfrak{A} \simeq \mathfrak{B}$ if p is such a morphism.

Recursive $\mathbb{L}$-equivalence is again an equivalence relation and we call the equivalence classes the $\mathbb{L}$-RETs. We write $\mathbb{L}\langle \mathfrak{A} \rangle$ for the $\mathbb{L}$-RET of $\mathfrak{A}$.

These $\mathbb{L}$-RETs are not quite the same as either version of the constructive order types (Crossley [1965], [1969]). What exactly the differences in the theory are is not presently known but most things are unaffected. In Crossley [1969] morphisms must always be extendable to recursively enumerable morphisms but here we only require that they can be extended to one-one partial recursive functions.

$\Omega(\mathbb{L})$ denotes the set of all $\mathbb{L}$-RETs. $A \in \Omega(\mathbb{L})$ is said to be *Dedekind* if the underlying set of some (or equivalently, all) $\mathfrak{A} \in A$ is (are) Dedekind in the $\mathbb{S}$-RET sense.

There are c Dedekind $\mathbb{L}$-RETs, each determined by some sub*set* of $\mathfrak{Q}$. However, we give for illustration a stronger version of this result.

Lemma 12.2. *Let $\mathfrak{A}$ be an infinite (linearly ordered) subordering of $\mathfrak{Q}$. There exist c Dedekind suborderings $\mathfrak{B}^f$ of $\mathfrak{Q}$ such that $\mathfrak{B}^f$ is isomorphic to $\mathfrak{A}$ and $\mathfrak{B}^f$ contains no infinite recursively enumerable subset.*

From this we shall immediately get

Corollary 12.3. *Given any countably infinite order type γ there exist c distinct Dedekind $\mathbb{L}$-RETs A^f such that (any representative of) each A^f has order type γ.* $\square$

Proof of the lemma. Two points $a, b \in \mathfrak{Q}$ are said to have the same type over a finite set $X \subseteq \mathfrak{Q}$ if the identity morphism $i: X \to X$ in $\mathbb{L}$ can be extended to a morphism of $\mathbb{L}$ from $X \cup \{a\}$ to $X \cup \{b\}$ (with the induced ordering).

Firstly, $\mathfrak{Q}$ has a subordering $\mathfrak{A}$ of type γ. Enumerate $\mathfrak{A}$ as $a^0, a^1, a^2, \ldots$. For each n, for each $i \in 2^n$ and for $j = 0, 1$ choose $\mathfrak{B}(i, j)$ to be an extension of $\mathfrak{B}(i)$ by one element $a^{i, j}$ such that $a^{i, 0}$ and $a^{i, 1}$ realize the same type over $\mathfrak{B}(i)$, $\mathfrak{B}(i, 0) \cap \mathfrak{B}(i, 1) = \mathfrak{B}(i)$ and $\mathfrak{B}(i, 0)$, $\mathfrak{B}(i, 1)$ are order isomorphic to $\{a^0, \ldots, a^n\}$ by isomorphisms extending the one from $\mathfrak{B}(i)$ to $\{a^0, \ldots, a^{n-1}\}$. Since $\mathfrak{Q}$ is infinite, dense and un-bordered we can choose such $a^{i, 0}$, $a^{i, 1}$. For each $f \in 2^\omega$ let $\mathfrak{B}^f = \bigcup \{\mathfrak{B}(i): i$ is an initial segment of $f\}$. By construction $\mathfrak{B}^f$ is of order type γ for each $f \in 2^\omega$. Now topologize $\mathcal{T} = \{\mathfrak{B}^f: f \in 2^\omega\}$ as a Cantor space by taking as basic open sets $U(i) = \{\mathfrak{B}^f: i \in 2^n$ for some n and i is an initial segment of $f\}$. Let $S_Z = \{\mathfrak{B}^f \in \mathcal{T}: \mathfrak{B}^f \supseteq Z\}$ then by the lemma below S_Z is nowhere dense if Z is infinite. Therefore $S = \bigcup \{S_Z: Z$ is an infinite recursively enumerable set$\}$ is a countable union of nowhere dense sets in a Cantor space. So S is first category. By Baire's category theorem, $\mathcal{T} - S$ is therefore second category of cardinality c. $\square$

Lemma 12.4. *If Z is infinite, S_Z is nowhere dense.*

Proof. Let $U(i)$ be a non-empty neighbourhood. We produce a non-empty subneighbourhood disjoint from S_Z. If $U(i) \cap S_Z$ is empty, $U(i)$ will do. If not choose $f \in U(i) \cap S_Z$ such that $i \subseteq f$ and $Z \subseteq \mathfrak{B}^f$. Then since Z is infinite we can choose $b \in Z$ such that $b \in \mathfrak{B}(j) - \mathfrak{B}(i)$ for some finite initial segment j of f. Given b choose $j \in 2^n$ with n as small as possible. If $j' = j|(n-1) \in 2^{n-1}$, then $i \subseteq j'$. Let $k \in 2^n$ be chosen so that $j' \subseteq k$ and $k(n-1) \neq j(n-1)$. Then $b \notin \mathfrak{B}(k)$ and in fact $b \notin \mathfrak{B}(k')$ for any $k' \supseteq k$. Hence $U(k)$ is the required neighbourhood. $\square$

13. Recursive Equivalence in a General Setting

The specific categories $\mathbb{S}$ and $\mathbb{L}$ of §§11, 12 are special cases where the theory of recursive equivalence can be developed. We now turn to a more general situation. From now on we shall consider categories obtained as follows. Let $\mathfrak{M}$ be a completely effective $\aleph_0$-categorical model. Let $\mathbb{C}$ be the category whose objects are all the algebraically closed subsystems of $\mathfrak{M}$ and whose morphisms are all the monomorphisms between those subsystems. We take as $°\mathbb{C}$ the full subcategory of finite systems of $\mathbb{C}$. So $\mathbb{C} = \mathbb{C}(\mathfrak{M})$ and is suitable.

Definition 13.1. Let $\mathfrak{A}, \mathfrak{B} \in \mathbb{C}$ then $\mathfrak{A}$ is said to be $\mathbb{C}$-*recursively equivalent* to $\mathfrak{B}$ if there is a one-one partial recursive function p such that

   (i) $\mathfrak{A} \subseteq \operatorname{dom} p$,

   (ii) p restricted to $\mathfrak{A}$ is a $\mathbb{C}$-isomorphism from $\mathfrak{A}$ to $\mathfrak{B}$.

   In this case we write $p: \mathfrak{A} \simeq \mathfrak{B}$.

Lemma 13.2. $\mathbb{C}$-*recursive equivalence is an equivalence relation.*

Proof. The relation is reflexive since the identity function is one-one partial recursive. If p is one-one and partial recursive then p^{-1} is well-defined, partial recursive and one-one where $p^{-1}(x) = y$ if, and only if, $p(y) = x$. So we have symmetry. If $p: \mathfrak{A} \simeq \mathfrak{B}$ and $q: \mathfrak{B} \simeq \mathfrak{C}$ then $q \circ p$ is a one-one partial recursive function and $q \circ p$ restricted to $|\mathfrak{A}|$ maps $\mathfrak{A}$ isomorphically onto $\mathfrak{C}$. Hence the relation is transitive. $\square$

By virtue of this lemma we write $\mathbb{C}\langle\mathfrak{A}\rangle$ for the equivalence class of $\mathfrak{A}$ under $\mathbb{C}$-recursive equivalence. We call these equivalence classes *effective $\mathbb{C}$-types.* (We always suppress reference to $\mathbb{C}$ where there is no ambiguity.) We write $\Omega(\mathbb{C})$ for the collection of all effective $\mathbb{C}$-types.

Lemma 13.3. *Suppose* $p: \mathfrak{A} \simeq \mathfrak{B}$ *then* $\hat{p}: \mathfrak{A} \simeq \mathfrak{B}$ *where* $\hat{p} = \bigcup \{q \subseteq p : q$ *is a finite morphism*$\}$.

Proof. Trivially $\hat{p} \subseteq p$. Now $\mathfrak{A} = \bigcup \{\mathfrak{A}' : \mathfrak{A}' \subseteq \mathfrak{A}\ \&\ \mathfrak{A}' \in °\mathbb{C}\}$. For any $x \in \mathfrak{A}$ let q^x be the restriction of p to $\operatorname{cl}\{x\}$. Since the algebraic closure of a finite set is finite $q^x \subseteq \hat{p}$. Hence $\hat{p}$ is a one-one function with domain containing $\mathfrak{A}$. $\hat{p}$ is partial recursive; for to enumerate the graph of $\hat{p}$, enumerate finite subsets q of the graph of p (which is possible since p is partial recursive). Then let $\hat{p}$ be the union of those finite subsets q such that $\operatorname{dom} q$, $\operatorname{ran} q$ are algebraically closed and q is a $\mathbb{C}$-morphism of $\operatorname{dom} q$ onto $\operatorname{ran} q$. This latter is effective since $\mathfrak{M}$ is fully effective.

Hence $\hat{p}$ is partial recursive. Finally since $\hat{p} \subseteq p$, p is structure preserving; and $\hat{p}$ maps $\mathfrak{A}$ onto $\mathfrak{B}$ since p does so. $\square$

Because of this lemma we may always assume that if $p: \mathfrak{A} \simeq \mathfrak{B}$ then p is of the form $\hat{p}$.

Definition 13.4. A one-one partial recursive function p is said to be a $\mathbb{C}$-*recursive equivalence* if $p = \hat{q}$ for some partial recursive one-one q.

Example 13.1. Let $\mathbb{V}$ be the category obtained from a countably infinite dimensional vector space over a finite field K. Since such a vector space is $\aleph_0$-categorical and fully effective, $\mathbb{V}$ is of the required sort. The $\mathbb{V}$-RETs are then the RETs of vector spaces (when the field K is finite) of Dekker [1969] except that our definition of recursive equivalence is wider as Dekker requires that morphisms always be extendable to recursively enumerable morphisms.

Example 13.2. Let $\mathfrak{B}$ be a fully effective model which is a countable atomless Boolean algebra. Then $\mathfrak{B}$ is $\aleph_0$-categorical and the associated category $\mathbb{B}$ contains isomorphs of all countable Boolean algebras. The finite $\mathbb{B}$-RETs are the (RETs of) Boolean algebras with 2^n elements for some $n \in E$ with $n > 0$.

Definition 13.5. A structure $\mathfrak{A}$ is said to be *effectively Dedekind finite* (Dedekind, for short) if $\mathfrak{A}$ contains no infinite recursively enumerable subset.

From §11 we have that $\mathfrak{A}$ is Dedekind if, and only if, there is no one-one partial recursive p which maps $\mathfrak{A}$ onto a proper subset of $\mathfrak{A}$.

Definition 13.6. A $\mathbb{C}$-type A is said to be *Dedekind* if some $\mathfrak{A} \in A$ is Dedekind.

"Some" may be replaced by "every" in Definition 13.6 for if $p: \mathfrak{A} \to X \subset \mathfrak{A}$ where p is one-one partial recursive and $q: \mathfrak{A} \simeq \mathfrak{B}$ then $q \circ p \circ q^{-1}: \mathfrak{B} \to q^{-1}(X) \subset \mathfrak{B}$ and $q \circ p \circ q^{-1}$ is partial recursive and one-one.

We write $\Lambda(\mathbb{C})$ for the collection of all Dedekind effective $\mathbb{C}$-types. Generally we shall call elements of $\Lambda(\mathbb{C})$ *Dedekind types*.

We observe that (i) if $\mathfrak{A}$ is Dedekind and $\mathfrak{B} \subseteq \mathfrak{A}$ then $\mathfrak{B}$ is Dedekind, (ii) if $\mathfrak{A}, \mathfrak{B} \subseteq \mathfrak{M}$ are Dedekind then $\mathfrak{A} \cap \mathfrak{B}$ is Dedekind and (iii) since $X \subseteq \mathrm{cl}(X)$ then $\mathrm{cl}(X)$ Dedekind implies X Dedekind.

The converse of (iii) is false. For consider a fully effective $\aleph_0$-categorical model which is a linear ordering of type $2 \cdot \eta$ (η copies of 2 where η is the order type of the rationals). Enumerate all the pairs with no element between them as $\{(a^n, b^n)\}_{n=0}^{\infty}$ with a, b recursive. (This is possible since these are the pairs satisfying the atom

$$\forall x (x \leq v_0 \vee v_1 \leq x) \,\&\, v_0 \neq v_1.)$$

Take I as an immune and co-immune set and let

$$A = \{a^n: n \in I\} \cup \{b^n: n \notin I\}.$$

Then since every element a satisfies

$$\exists! {}_1 \forall x ((x < a \rightarrow x \leq y) \vee (a < x \rightarrow y \leq x))$$

$\mathrm{cl}(A)$ is the universe of the model which is not Dedekind.

14. Existence of Dedekind Types

We show that in any suitable category $\mathbb{C}$ there exist continuum many, that is, $c = 2^{\aleph_0}$ infinite Dedekind $\mathbb{C}$-types. In fact we prove a stronger result.

Theorem 14.1. *Let $\mathbb{C}$ be a suitable category. Then, for any (classical) isomorphism type of an object in $\mathbb{C}$, there exist continuum many distinct Dedekind $\mathbb{C}$-types of that isomorphism type.*

The proof is just the generalization of that for Lemma 12.2 using Lemmata 5.4, 5.5 and 5.9 to ensure that the method applies to suitable categories.

Proof. Let $\mathfrak{A}$ be an infinite object in $\mathbb{C}$. By Lemma 5.9 there exists a sequence $\mathfrak{A}^0 \subset \mathfrak{A}^1 \subset \cdots$ of objects $\mathfrak{A}^n \in {}^\circ\mathbb{C}$ such that $\bigcup \mathfrak{A}^i = \mathfrak{A}$. We write 2^n for the set of functions from n into 2. If $i \in 2^m$ and $j \in 2^n$ we let $i \wedge j$ be the largest common initial segment of (the sequences) i and j. Now by Lemma 5.5 we can inductively (under the inclusion ordering) choose $\mathfrak{B}^i \in {}^\circ\mathbb{C}$ and an isomorphism $p^i \in {}^\circ\mathbb{C}$ where, if $i \in 2^n$, $p^i: \mathfrak{B}^i \simeq \mathfrak{A}^{n+1}$ such that $\mathfrak{B}^i \cap \mathfrak{B}^j = \mathfrak{B}^{i \wedge j}$ and $p^i \cap p^j = p^{i \wedge j}$. For $f \in 2^\omega$ let $\mathfrak{B}^f = \bigcup \{\mathfrak{B}^i: i \text{ is an initial segment of } f\}$. Then $f \leadsto \mathfrak{B}^f$ maps 2^ω one-one onto $\{\mathfrak{B}^f: f \in 2^\omega\}$. Each $\mathfrak{B}^f$ is $\mathbb{C}$-isomorphic to $\mathfrak{A}$ by $\bigcup \{p^i: i \text{ is an initial segment of } f\}$. Topologize 2^ω with the usual Cantor set topology. Then the basic open neighbourhoods are the sets $U(i) = \{f \in 2^\omega: i \subseteq f\}$ for $i \in 2^n$ for some n.

If Z is an infinite subset of $\mathfrak{M}$, then

$$S_Z = \{f: \mathfrak{B}^f \supseteq Z\}$$

is nowhere dense by the same proof as for Lemma 12.4. By the Baire category theorem the union of all the countably infinitely many S_Z with Z an infinite recursively enumerable set has a complement of cardinal $c = 2^{\aleph_0}$. Since each $\mathbb{C}$-RET contains only countably many representatives the elements in the complement give rise to $2^{\aleph_0}$ distinct $\mathbb{C}$-RETs all of the isomorphism type of the original $\mathfrak{A}$. $\quad\square$

15. Partial Recursive Combinatorial Functors

Let $^{re}\mathbb{C}$ be the subcategory of $\mathbb{C} = \mathbb{C}(\mathfrak{M})$ whose objects are the recursively enumerable algebraically closed sets of $\mathfrak{M}$ and whose morphisms are those elementary monomorphisms between objects which are partial recursive as maps. (A map is partial recursive if its domain, graph, and codomain are all recursively enumerable.)

The assumption $\mathbb{C} = \mathbb{C}(\mathfrak{M})$ is suitable implies that $^{\circ}\mathbb{C}$ has a Gödel numbering effectively assigning a natural number $\ulcorner p \urcorner$ to each morphism p of $^{\circ}\mathbb{C}$, so that the cardinalities of dom p, codom p, graph p can be effectively obtained from $\ulcorner p \urcorner$. We call $\ulcorner p \urcorner$ an *explicit index* of p. Also $^{re}\mathbb{C}$ inherits a Gödel numbering from Kleene's standard enumeration of all recursively enumerable sets and all partial recursive functions. We call these *standard indices*.

We wish to define the notion of a partial recursive combinatorial functor. This definition comes in two parts because the notion of combinatorial functor comes in two parts. One part is the functorial assignment of maps to maps. The other part concerns $F^{\leftarrow} x$, tracing an element x of an image object to the precise finite preimage object whence it arises.

Definition 15.1. Let $\mathbb{C} = \mathbb{C}(\mathfrak{M})$ and $\mathbb{C}' = \mathbb{C}(\mathfrak{M}')$. A combinatorial functor $F: \mathbb{C} \to \mathbb{C}'$ is said to be *partial recursive* if F maps $^{\circ}\mathbb{C}$ into $^{re}\mathbb{C}'$ and: (i) There is a general recursive function assigning to the explicit index $\ulcorner p \urcorner$ of each p in $^{\circ}\mathbb{C}$ a standard index $\ulcorner Fp \urcorner$ of Fp in $^{re}\mathbb{C}'$, (ii) The map from $x \in \mathfrak{M}'$ to the explicit index $\ulcorner F^{\leftarrow} x \urcorner$ of $F^{\leftarrow} x$ in $^{\circ}\mathbb{C}$ is partial recursive.

Note that (i) is the requirement that F be a partial recursive functional on morphisms.

Theorem 15.2. *Let* $F: \mathbb{C} \to \mathbb{C}'$ *be a partial recursive combinatorial functor. Then* F *induces a function* F *from* $\mathbb{C}$-*RETs to* $\mathbb{C}'$-*RETs given by* $\mathsf{F}\langle \mathfrak{A} \rangle = \langle F\mathfrak{A} \rangle$ *for objects* $\mathfrak{A}$ *of* $\mathbb{C}$.

Proof. Suppose $\mathfrak{A}$, $\mathfrak{B}$ are objects of $\mathbb{C}$ and $p: \mathfrak{A} \simeq \mathfrak{B}$. By assumption p is 1-1 and partial recursive. Let $r = \bigcup \{\text{graph } F(q): q \in ^{\circ}\mathbb{C} \text{ and } q \subseteq p\}$. We prove first that r is a 1-1 function. It suffices to show that for $q, q' \in ^{\circ}\mathbb{C}$, if $q, q' \subseteq p$, then $F(q) \cap F(q')$ is defined. This follows from Theorem 3.12 since $q, q' \subseteq p$ and p 1-1 imply $q \cap q'$ is defined.

Next we show r is partial recursive. Since p is partial recursive, the set of $\ulcorner q \urcorner$ with q in $^{\circ}\mathbb{C}$ and $q \subseteq p$, is recursively enumerable. Since F is partial recursive, the set of $\ulcorner F(q) \urcorner$ with q in $^{\circ}\mathbb{C}$, $q \subseteq p$ is also recursively enumerable. So r is the union of a recursively enumerable set of partial recursive functions, and is partial recursive. Last, we show $r: F(\mathfrak{A}) \simeq F(\mathfrak{B})$. But r suitably restricted to have domain $F(\mathfrak{A})$ and codomain $F(\mathfrak{B})$ is exactly the map $F(p')$, where p' is the restriction of p to domain $\mathfrak{A}$ and codomain $\mathfrak{B}$. This is obvious when $F(p')$ is thought of as the directed union of all $F(p'')$ with $p'' \subseteq p'$, $p'' \in {}^{\circ}\mathfrak{C}$. $\square$

Theorem 15.3. *Partial recursive combinatorial functors are closed under composition.*

Proof. Let $F: \mathfrak{C} \to \mathfrak{C}'$, let $G: \mathfrak{C}' \to \mathfrak{C}''$. If $p \in {}^{\circ}\mathfrak{C}$, then $G(F(p)) = \bigcup \{G(q): q \in {}^{\circ}\mathfrak{C}' \& q \subseteq F p\}$. From the index $\ulcorner p \urcorner$ of p find the standard index $\ulcorner F p \urcorner$ of $F p$. Use this to effectively enumerate all $\ulcorner q \urcorner$ with $q \in {}^{\circ}\mathfrak{C}'$, $q \subseteq F p$. For each such $\ulcorner q \urcorner$, effectively find the standard index $\ulcorner G q \urcorner$ of $G q$. This process recursively enumerates all the $\ulcorner G q \urcorner$, and yields a standard index $\ulcorner G(F(p)) \urcorner$ of the union of all such $G q$. This verifies (i). As for (ii), first note

$$x \in G(F(\mathfrak{A})) \leftrightarrow G^{\leftarrow} x \subseteq F \mathfrak{A} \leftrightarrow \bigcup \{F^{\leftarrow} y: y \in G^{\leftarrow} x\} \subseteq \mathfrak{A}.$$

So

$$(G \circ F)^{\leftarrow} x = \mathrm{cl} \bigcup \{F^{\leftarrow} y: y \in G^{\leftarrow} x\}.$$

Given x, compute the explicit index $\ulcorner F^{\leftarrow} x \urcorner$ and use this to produce a finished list $y_1, \dots, y_k$ of members of $F^{\leftarrow} x$. For each y_i, compute the explicit index $\ulcorner F^{\leftarrow} y_i \urcorner$. If all this succeeds and is finished, use the effectiveness of closure in $\mathfrak{C}$ to compute an explicit index of $\mathrm{cl} \bigcup \{F^{\leftarrow} y: y \in G^{\leftarrow} x\}$. This is $\ulcorner (G \circ F)^{\leftarrow} x \urcorner$ as required. $\square$

Theorem 15.4. *Let $F: \mathfrak{C} \to \mathfrak{C}'$ be a partial recursive combinatorial functor. Suppose F is finitary (that is, F maps ${}^{\circ}\mathfrak{C}$ to ${}^{\circ}\mathfrak{C}'$). Then $\mathfrak{A}$ is Dedekind implies $F \mathfrak{A}$ is Dedekind.*

Proof. Suppose $\mathfrak{A}$ is Dedekind and $X \subseteq F(\mathfrak{A})$ is recursively enumerable. Since F is combinatorial, $Y = \bigcup \{F^{\leftarrow} x: x \in X\} \subseteq \mathfrak{A}$. Since F is partial recursive, condition (ii) of Definition 15.1 implies that Y is recursively enumerable. Therefore Y is finite. So $\mathrm{cl}\, Y$ is finite too. Since F is finitary, $F(\mathrm{cl}\, Y)$ is finite. But $X \subseteq F(\mathrm{cl}\, Y)$ because $\bigcup \{F^{\leftarrow} x: x \in X\} \subseteq \mathrm{cl}\, Y$. So X is finite. $\square$

The assumption that F is finitary is essential for Theorem 15.4, for consider the constant functor on $\mathfrak{C}(\mathfrak{M})$ to $\mathfrak{C}(\mathfrak{M}')$ with value $\mathfrak{M}'$.

For sufficiently effective finitary functors, part (ii) of Definition 15.1 is redundant.

Definition 15.5. A combinatorial functor $F: \mathfrak{C} \to \mathfrak{C}'$ is said to be *finitary recursive* if F maps ${}^{\circ}\mathfrak{C}$ to ${}^{\circ}\mathfrak{C}'$ and there is a general recursive function assigning to the explicit index $\ulcorner p \urcorner$ of each p in ${}^{\circ}\mathfrak{C}$ an *explicit* index $\ulcorner F p \urcorner$ of $F p$ in ${}^{\circ}\mathfrak{C}'$.

Theorem 15.6. *Finitary recursive combinatorial functors are partial recursive combinatorial functors.*

Proof. Remember that identity maps and objects are identified. By assumption $\bigcup\{F\mathfrak{A}: \mathfrak{A}\in{}^\circ\mathbb{C}\}$ is recursively enumerable. Suppose x is in this set. Then we can find an $\mathfrak{A}\in{}^\circ\mathbb{C}$ with $x\in F\mathfrak{A}$ by enumerating the values of F on ${}^\circ\mathbb{C}$. For this $\mathfrak{A}$ we can explicitly list all $\mathfrak{A}'\in{}^\circ\mathbb{C}$, $\mathfrak{A}'\subseteq\mathfrak{A}$ since $\mathbb{C}$ is suitable, and since F is finitary recursive we can find the explicit index $\ulcorner F\mathfrak{A}'\urcorner$ for each such $\mathfrak{A}'$. Among these $\mathfrak{A}'$ we can find the smallest, $\mathfrak{A}''$, with $x\in F\mathfrak{A}''$. Then $\ulcorner F^{\leftarrow} x\urcorner = \ulcorner\mathfrak{A}''\urcorner$.   □

Theorem 15.7. *Finitary recursive combinatorial functors are closed under composition.*

Proof. Let $F: \mathbb{C}\to\mathbb{C}'$, $G: \mathbb{C}'\to\mathbb{C}''$ be such. Then from the explicit index $\ulcorner\mathfrak{A}\urcorner$, for $\mathfrak{A}\in{}^\circ\mathbb{C}$, first compute the explicit index $\ulcorner F\mathfrak{A}\urcorner$, then compute the explicit index $\ulcorner G\circ F(\mathfrak{A})\urcorner$. The theorem now follows from Theorem 3.2.   □

16. Partial Recursive Strict Combinatorial Functors

We assume here that all categories $\mathbb{C}(\mathfrak{M})$ are suitable and that $\mathfrak{M}$ has degree 1 and dimension.

Definition 16.1. A precombinatorial operator $G: \mathrm{Ob}\,{}^\circ\mathbb{C}(\mathfrak{M})\to\mathscr{P}(\mathfrak{M}')$ is said to be *recursive* if there is a general recursive function $\ulcorner G\urcorner$ which when applied to an explicit index $\ulcorner\mathfrak{A}\urcorner$ of $\mathfrak{A}$ in ${}^\circ\mathbb{C}$ yields a standard index $\ulcorner G\mathfrak{A}\urcorner$ of the recursively enumerable set $G\mathfrak{A}$.

Theorem 16.2. *Let $F: \mathbb{C}\to\mathbb{C}'$ be a strict combinatorial functor induced by a recursive precombinatorial operator as in Theorem 7.4. Then F is a partial recursive strict combinatorial functor.*

·*Proof.* We effectivize the proof of Theorem 7.4. First we show how to compute a standard index $\ulcorner Fp\urcorner$ of Fp from an explicit index $\ulcorner p\urcorner$ of $p: \mathfrak{A}\to\mathfrak{B}$ in ${}^\circ\mathbb{C}$. Now let G be a recursive precombinatorial operator inducing F. We can simultaneously effectively enumerate all $G\mathfrak{A}$ for $\mathfrak{A}\in{}^\circ\mathbb{C}$, so the function f matching $\bigcup\{G\mathfrak{A}': \mathfrak{A}'\subseteq\mathfrak{A}\ \&\ \mathfrak{A}'\in{}^\circ\mathbb{C}\}$ with $\bigcup\{G\mathfrak{B}': \mathfrak{B}'\subseteq\mathfrak{B}\ \&\ \mathfrak{B}'\in{}^\circ\mathbb{C}\}$ is one-one and partial recursive. The proof of Lemma 5.2 is effective, since by our assumption $\mathbb{C}(\mathfrak{M})$ is suitable, and so $\mathrm{cl}\,f$ is also one-one partial recursive and uniformly so in the explicit index $\ulcorner p\urcorner$ of p, since $k=1$ as $\mathfrak{M}$ has degree 1. Hence we can indeed find a standard index $\ulcorner Fp\urcorner$ of Fp from the explicit index $\ulcorner p\urcorner$ of p.

Now we show condition (ii) of Definition 15.2 is satisfied. By the definition of F from G, $F^{\leftarrow} x=\mathrm{cl}(\mathfrak{A}^1\cup\cdots\cup\mathfrak{A}^k)$ where $\mathfrak{A}^1,\ldots,\mathfrak{A}^k$ are the finitely many objects of ${}^\circ\mathbb{C}$ such that $\mathrm{supp}(x)\cap G\mathfrak{A}^i\neq\emptyset$. Since G is recursive and precombinatorial we can recursively enumerate the independent set $B=\bigcup\{G\mathfrak{A}: \mathfrak{A}\in{}^\circ\mathbb{C}\}$. We use this enumeration to compute the $\mathfrak{A}^i$. First find an atom ϕ and elements $b_1, \ldots, b_k$ of B such that $\mathfrak{M}' \models \phi(x, b_1, \ldots, b_k)$. Now check through the finite number of subsets of $\{b_1, \ldots, b_k\}$ and the finite number of atoms which imply ϕ to compute the support of x. Then for each $b \in \mathrm{supp}(x)$ we can compute $\ulcorner \mathfrak{A} \urcorner$ for $b \in G\mathfrak{A}$. These $\mathfrak{A}$ are $\mathfrak{A}^1, \ldots, \mathfrak{A}^k$. Using once more the fact that $\mathfrak{M}$ is suitable we can then compute an explicit index of $\mathrm{cl}(\mathfrak{A}^1 \cup \cdots \cup \mathfrak{A}^k)$, that is, we can compute $\ulcorner F^- x \urcorner$. $\quad\square$

Corollary 16.3. *If $\mathbb{C}(\mathfrak{M})$ is suitable and $\mathfrak{M}$ has degree 1, then the closure functor V obtained from a recursively enumerable basis for $\mathfrak{M}$ is partial recursive combinatorial.* $\quad\square$

Theorem 16.4. *Let $F^1, F^2 \colon \mathbb{C} \to \mathbb{C}'$ be strict combinatorial functors induced by recursive precombinatorial operators G^1, G^2. Suppose that for $\mathfrak{A} \in {}^\circ\mathbb{C}$, $\dim G^1 \mathfrak{A} = \dim G^2 \mathfrak{A}$ (or equivalently, that F^1, F^2 induce the same combinatorial series). Then F^1, F^2 induce the same function from $\mathbb{C}$-RETs to $\mathbb{C}'$-RETs.*

Proof. We effectivize the proof of Theorem 7.6 (compare the proof of Theorem 16.2). Using the simultaneous effective enumerations without repetition of the $G^i \mathfrak{A}$ we see that the τ matching $G^1 \mathfrak{A}$ with $G^2 \mathfrak{A}$ for all $\mathfrak{A} \in {}^\circ\mathbb{C}$ is one-one and partial recursive. Hence the restriction $\tau_{\mathfrak{A}} \colon F^1 \mathfrak{A} \to F^2 \mathfrak{A}$ is an isomorphism which can be extended to a one-one partial recursive function. Hence $\langle F^1 \mathfrak{A} \rangle = \langle F^2 \mathfrak{A} \rangle$ for all $\mathfrak{A}$. $\quad\square$

Theorem 16.5. *Let $V^i \colon \mathbb{S} \to \mathbb{C}^i$ be partial recursive closure functors. Let $F \colon \mathbb{C}^1 \to \mathbb{C}^2$ be a strict partial recursive combinatorial functor induced by a recursive precombinatorial operator. Then there exists a (strict) partial recursive combinatorial functor $H \colon \mathbb{S} \to \mathbb{S}$ induced by a recursive precombinatorial operator such that the following diagram commutes.*

$$
\begin{array}{ccc}
\Omega(\mathbb{C}^1) & \xrightarrow{\ F\ } & \Omega(\mathbb{C}^2) \\[2pt]
{\scriptstyle V^1}\Big\uparrow & & \Big\uparrow{\scriptstyle V^2} \\[2pt]
\Omega(\mathbb{S}) & \xrightarrow[\ H\ ]{} & \Omega(\mathbb{S})
\end{array}
$$

Proof. Effectivize the proofs of Lemmata 7.10 and 7.11 following the pattern of the proofs of Theorems 16.2 and 16.4 above. $\quad\square$

Similarly one obtains the following two theorems.

Theorem 16.6. *For a function $f \colon \times^k E \to E$ the following are equivalent*

(i) *f is recursive and strictly $\mathbb{C}$-combinatorial,*

(ii) *f is induced by a finitary recursive strict combinatorial functor.*

(iii) *the Stirling coefficient function of f is recursive.* $\quad\square$

Theorem 16.7. *Strict recursive $\mathbb{C}$-combinatorial functions are closed under composition.* $\quad\square$

Now $\mathbb{S}$-combinatorial *series* with recursive Stirling coefficient functions are *not* closed under composition. This is what led to partial recursive, as opposed to recursive, combinatorial functors in the first place.

Example 16.1. (Cf. Nerode [1966] for further details.) Let $R \subseteq E \times E$ be a recursive relation such that $R \cap (E \times \{0\}) = \emptyset$ and $S = \{x : \exists y ((x,y) \in R)\}$ is recursively enumerable but not recursive. Let

$$f(m, n) = \sum c^{ij} \binom{m}{i} \binom{n}{j}$$

where

$$c^{ij} = 1 \quad \text{if } (i, j) \in R,$$

$$= 0 \quad \text{otherwise}.$$

Clearly f is recursive.

Let $g(n) = \aleph_0 \left(= \aleph_0 \binom{n}{0} \right)$. Then $f(m, g(n)) = f(m, \aleph_0) = \sum c^{ij} \binom{m}{i} \binom{\aleph_0}{j}$.
Now $\binom{\aleph_0}{j} = \aleph_0$ if $j > 0$ and also $i \in S$ if, and only if, $\exists j ((i, j) \in R)$ by definition, so

$$f(m, g(n)) = \sum_{i \in S} \binom{m}{i} \aleph_0.$$

But this series not have a recursive coefficient function for the series is $\sum \aleph_0 d(i) \binom{m}{i}$ where $d(i)$ is the characteristic function of S which, by the choice of S, is not recursive.

We therefore inquire which coefficient functions arise from partial recursive strict combinatorial functors induced by recursive precombinatorial operators G. The only significant requirements are that $G\mathfrak{A}$ be uniformly recursively enumerable in $\mathfrak{A}$ for $\mathfrak{A} \in {}^\circ\mathbb{C}$ and that if $\mathfrak{A}$ and $\mathfrak{B}$ have the same dimension then $G\mathfrak{A}$ and $G\mathfrak{B}$ have the same cardinality.

Definition 16.8. A function $c : \times^k E \to E^+ = E \cup \{\aleph_0\}$ is said to be an $R\uparrow$ *function* (or a limit of a monotone increasing sequence of recursive functions) if there is a recursive sequence of recursive functions $f^n : \times^k E \to E$ such that for all i in $\times^k E$, $f^0(i) \le f^1(i) \le \cdots$ and $c(i)$ is the supremum in E^+ of $\{f^n(i) : n = 0, 1, 2, \ldots\}$.

A $\mathbb{C}$-combinatorial series $g(x) = \sum c(i) \begin{bmatrix} x \\ i \end{bmatrix}$ is said to be an $R\uparrow$ *series* if $c(i)$ is an $R\uparrow$ function.

Clearly the $R\uparrow$ series (properly) include the series with recursive coefficient functions.

Lemma 16.9. *A function* $c : \times^k E \to E^+$ *is* $R\uparrow$ *if, and only if, there is a recursively enumerable family of recursively enumerable sets* $\{B^i : i \in E\}$ *such that for each i, B^i has $c(i)$ elements.*

Proof. Suppose that the functions f^n: $\times^k E \to E$ constitute a monotone increasing recursive sequence of recursive functions with $c(i) = \sup\{f^n(i): n \in E\}$. Let $B^i = \{n: \exists j(f^j(i) > n)\}$. Then B^i consists of the $c(i)$ elements $0, 1, \ldots, m-1$ where $m = \sup\{f^n(i): n \in E\}$ and clearly the B^i constitute a recursively enumerable family of recursively enumerable sets.

Conversely if $\{B^i: i \in \times^k E\}$ is given then we can enumerate the elements of the B^i so that only a finite number have been enumerated at any finite stage. Let $f^n(i)$ be the number of elements of B^i enumerated by stage n. Then the sequence $\{f^n(i): n = 0, 1, 2, \ldots\}$ clearly has the required properties. $\square$

Theorem 16.10. *A $\mathbb{C}$-combinatorial series is $R\uparrow$ if, and only if, it is induced by a partial recursive strict $\mathbb{C}$-combinatorial functor (which in turn is induced by a recursive precombinatorial operator).*

Proof. Effectivize Lemma 7.2 and then apply Theorem 16.2. $\square$

Corollary 16.11. *$R\uparrow$ $\mathbb{C}$-combinatorial series are closed under composition.*

Proof. Strict $\mathbb{C}$-combinatorial functors are closed under composition so all that is required in addition is an effective version of Theorem 7.9 but with a computation of G^3 from the precombinatorial operators inducing F^1 and F^2. $\square$

Corollary 16.12. *Let F, G be $R\uparrow$ strict $\mathbb{C}$-combinatorial functors inducing $R\uparrow$ $\mathbb{C}$-combinatorial series $F^\#$, $G^\#$ then $(F \circ G)^\# = F^\# \circ G^\#$ where $(F \circ G)^\#$ is the series induced by $F \circ G$.* $\square$

Part V. Identities

17. The Strong Topology

We now begin our investigations of properties which hold for "almost all" $\mathfrak{A}$ in $\mathbb{C}$ or which hold "uniformly on a neighbourhood". Nerode's combinatorial series [1966] generalized Myhill's combinatorial functions [1958]. We generalize combinatorial series from $\mathbb{S}$ to other categories but first we need to do some topology.

$\mathbb{C}$ (with or without superscripts) will be the category of algebraically closed subsystems of a fully effective $\aleph_0$-categorical model $\mathfrak{M}$ with monomorphisms as the morphisms. We assume the universe $|\mathfrak{M}|$ of $\mathfrak{M}$ is the natural numbers E. We write $\mathfrak{A}$ for $|\mathfrak{A}|$ where no confusion arises.

Lemma 17.1. *The sets*

$$U(\mathfrak{A}, N) = \{\mathfrak{A}' \in \mathbb{C}: \mathfrak{A} \subseteq \mathfrak{A}' \,\&\, \mathfrak{A}' \cap N = \emptyset\},$$

for $\mathfrak{A} \in {}^\circ\mathbb{C}$ and N a finite subset of E, form a basis for a topology, $\mathcal{T}(\mathbb{C})$ on the objects of $\mathbb{C}$.

Proof. We only need to show the intersection of two basic open sets is again a basic open set.

$$U(\mathfrak{A}, N) \cap U(\mathfrak{B}, M) = \{\mathbb{C} \in \mathbb{C}: \mathfrak{A} \cup \mathfrak{B} \subseteq \mathbb{C} \,\&\, \mathbb{C} \cap (M \cup N) = \emptyset\}$$
$$= U(\mathfrak{D}, M \cup N)$$

where $\mathfrak{D} = \mathrm{cl}(\mathfrak{A} \cup \mathfrak{B})$ since $\mathfrak{A} \cup \mathfrak{B} \subseteq \mathbb{C}$ and $\mathbb{C}$ algebraically closed implies $\mathfrak{D} \subseteq \mathbb{C}$. $\square$

We call the topology of $\mathcal{T}(\mathbb{C})$ the *strong* topology on $\mathbb{C}$ to distinguish it from the weak topology $\mathcal{T}^\circ$ of § 1.

Remark. The exact relation between the weak topology of § 1 and the present strong topology for the categories $\mathbb{C}$ we are considering now is as follows. The basic sets for the weak topology $\mathcal{T}^\circ$ are the sets $U(\mathfrak{A}, \emptyset)$ for $\mathfrak{A} \in {}^\circ\mathbb{C}$. These sets with their complements (in $\mathbb{C}$) form a subbase (Kelley [1955], p. 48) for the strong topology.

For if $U(\mathfrak{A}, N)$ is non-empty, then $|\mathfrak{A}| \cap N$ is empty and $N = \{n_1, \ldots, n_k\}$. Put $U^i = U(\text{cl}\{n_i\}, \emptyset)$ and observe

$$U(\mathfrak{A}, N) = U(\mathfrak{A}, \emptyset) \cap \bigcap_{i=1}^{i=k} |\mathfrak{M}| - U^i.$$

As usual a subset D of $\mathbb{C}$ is said to be *dense* in (the strong topology of) $\mathbb{C}$ if D intersects every open set, or equivalently, $D \cap U(\mathfrak{A}, N) \neq \emptyset$ for every basic neighbourhood $U(\mathfrak{A}, N)$.

A type X in $\Omega(\mathbb{C})$ is said to be finite if $X = \langle \mathfrak{X} \rangle$ for some $\mathfrak{X} \in {}^\circ\mathbb{C}$. For types X, Y we write $X \subseteq Y$ if there exist $\mathfrak{X} \in X$, $\mathfrak{Y} \in Y$ with $\mathfrak{X} \subseteq \mathfrak{Y}$. (There will be no confusion since $X \subseteq Y$ set-theoretically means $X = Y$ as both X and Y are equivalence classes.)

Lemma 17.2. *A type* $Y \in \Omega(\mathbb{C})$ *is dense in the weak topology if, and only if, for any finite type* $X \in \Omega(\mathbb{C})$, $X \subseteq Y$, *if, and only if,* Y *is dense in the strong topology.*

Proof. Let X be a finite type and $\mathfrak{X} \in X$, then $U(\mathfrak{X}, \emptyset)$ is a basic neighbourhood. If Y is dense in the weak topology, there exists $\mathfrak{Y} \in Y \cap U(\mathfrak{X}, \emptyset)$ but then $\mathfrak{X} \subseteq \mathfrak{Y}$ and $X \subseteq Y$. Now let $U(\mathfrak{X}, N)$ be a basic neighbourhood and let $X = \langle \mathfrak{X} \rangle$. For some $\mathfrak{X}' \in X$ there exists $\mathfrak{Y}' \supseteq \mathfrak{X}'$ with $\mathfrak{Y}' \in Y$. By the Duplication Lemma (5.5) there exists $\mathfrak{Y} \simeq \mathfrak{Y}'$ such that $\mathfrak{Y} \supseteq \mathfrak{X}$ and $\mathfrak{Y} \cap N = \emptyset$. But then $\mathfrak{Y} \in U(\mathfrak{X}, N)$ so Y is dense in the strong topology. Trivially, if Y is dense in the strong topology, it is dense in the weak. $\quad\square$

Lemma 17.3. *The dense types are co-meagre in the strong topology.*

Proof. There is only a countable set of finite types so it suffices to show that for X finite $\{\mathfrak{A} : \mathfrak{A} \in \mathbb{C} \& X \nsubseteq \langle \mathfrak{A} \rangle\}$ is nowhere dense. Let $\mathfrak{X} \in X$ and let $U(\mathfrak{Y}, N)$ be a non-empty basic neighbourhood. By the Duplication Lemma (5.5) there exists a finite $\mathfrak{Z}$ containing $\mathfrak{Y}$ such that $\text{cl}(\mathfrak{X} \cup \mathfrak{Y})$ is isomorphic to $\mathfrak{Z}$ and $\mathfrak{Z} \cap N = \emptyset$. Then $U(\mathfrak{Z}, N)$ is a sub-neighbourhood of $U(\mathfrak{Y}, N)$ and $\langle \mathfrak{A} \rangle \supseteq X$ for all $\mathfrak{A} \in U(\mathfrak{Z}, N)$, so the proof is complete. $\quad\square$

Lemma 17.3 shows that there are always plenty of dense types. In the case of categories with dimension, all infinite types are dense. If X is a finite type then we define dim $X = \dim \mathfrak{X}$ for any $\mathfrak{X} \in X$. dim X is well-defined here since $\mathfrak{A} \simeq \mathfrak{B}$ implies $\dim \mathfrak{A} = \dim \mathfrak{B}$.

Lemma 17.4. *Let* $\mathbb{C}$ *have dimension then all infinite* $\mathbb{C}$*-types are dense.*

Proof. Let X be a finite type and dim $X = n$. Let Y be an arbitrary infinite type and $\mathfrak{Y} \in Y$. Let $y^1, \ldots, y^n$ be n algebraically independent elements in $\mathfrak{Y}$. Such exist since $\mathfrak{Y}$ is infinite and the algebraic closure of a finite set is finite. Now $\mathfrak{Y}' = \text{cl}\{y^1, \ldots, y^n\} \subseteq \mathfrak{Y}$ and $\mathfrak{Y}'$ has dimension n. Hence $X \subseteq Y$. $\quad\square$

Even if all infinite types are dense there may not be a dimension function. For example, consider the category $\mathbb{L}$ of linear orderings. Any infinite linearly ordered set contains linearly ordered finite subsets of all finite order types. The case of Boolean algebras is similar: Every infinite Boolean algebra contains finite Boolean algebras of all possible finite types (of cardinalities 2^n for $n \in E$).

18. Extending Identities to Dedekind Dense Types

Definition 18.1. Let $F, G: \mathbb{C}^1 \to \mathbb{C}^2$ be combinatorial functors then F, G are said to be *uniformly effectively equivalent* on a neighbourhood $U = U(\mathfrak{A}^1, N)$ if there is a one-one partial recursive function p such that for all $\mathfrak{A} \in U$, p is defined on $F(\mathfrak{A})$ and $p|F(\mathfrak{A})$ is a $\mathbb{C}^2$-morphism of $F(\mathfrak{A})$ onto $G(\mathfrak{A})$.

Theorem 18.2. *Suppose $F, G: \mathbb{C}^1 \to \mathbb{C}^2$ are partial recursive combinatorial functors inducing functions F, G from $\mathbb{C}^1$-types to $\mathbb{C}^2$-types. Then the following conditions are equivalent:*

(i) $\mathsf{F}(\mathsf{X}) = \mathsf{G}(\mathsf{X})$ *for all dense types X in $\Omega(\mathbb{C}^1)$,*

(ii) $\mathsf{F}(\mathsf{X}) = \mathsf{G}(\mathsf{X})$ *for all Dedekind dense types X in $\Lambda(\mathbb{C}^1)$,*

(iii) *F and G are uniformly effectively equivalent on some non-empty neighbourhood.*

Proof. Trivially (i) implies (ii).

Every neighbourhood contains a representative of every dense type so (iii) implies (i).

Suppose (iii) is false. We shall apply the Baire category theorem to $\mathbb{C}^1$ to show that the types X such that $\mathsf{F}(\mathsf{X}) = \mathsf{G}(\mathsf{X})$ are meagre. Now the non-dense types are meagre by Lemma 17.3. Also by the proof of Theorem 14.1 the non-Dedekind types are meagre. Hence the set of Dedekind dense types X such that $\mathsf{F}(\mathsf{X}) \neq \mathsf{G}(\mathsf{X})$ is co-meagre (and in particular non-empty since $\mathbb{C}^1$ is a Cantor space).

So now we show that $\{\mathfrak{A}: \mathsf{F}\langle\mathfrak{A}\rangle = \mathsf{G}\langle\mathfrak{A}\rangle\}$ is first category. Since there are only countably many one-one partial recursive functions it suffices to show that for any partial recursive p the set of $\mathfrak{A}$ in $\mathbb{C}^1$ such that $p|F\mathfrak{A}$ is a $\mathbb{C}^2$-morphism of $F\mathfrak{A}$ onto $G\mathfrak{A}$ is nowhere dense. Let $U(\mathfrak{A}^1, N)$ be a non-empty neighbourhood in $\mathbb{C}^1$. Since (iii) fails for $U(\mathfrak{A}^1, N)$ and p (see Definition 18.1) † one of the following holds for some $\mathfrak{A}^2 \in U(\mathfrak{A}^1, N)$.

(1) p is not defined on all of $F\mathfrak{A}^2$,

(2) p^{-1} is not defined on all of $G\mathfrak{A}^2$,

(3) for some $a \in F\mathfrak{A}^2$, $p(a) \notin G\mathfrak{A}^2$,

(4) for some $a \in G\mathfrak{A}^2$, $p^{-1}(a) \notin F\mathfrak{A}^2$,

(5) $p|F\mathfrak{A}^2$ is not a morphism of $\mathbb{C}^2$ but (1)–(4) all fail.

(If (1)–(5) all fail then F, G are uniformly effectively equivalent by p on $U(\mathfrak{A}^1, N)$.)

Since F, G are induced by ${}^\circ F$, ${}^\circ G \colon {}^\circ\mathbb{C}^1 \to \mathbb{C}^2$ we may assume $\mathfrak{A}^2$ is finite and therefore in ${}^\circ\mathbb{C}^1$.

Cases (1) and (2), and (3) and (4) are symmetrical so we treat only (1), (3) and (5).

In case (1), $U(\mathfrak{A}^2, N)$ is the required non-empty subneighbourhood for: if $\mathfrak{A} \in U(\mathfrak{A}^2, N)$ then $F\mathfrak{A} \supseteq F\mathfrak{A}^2$, but p is not defined on all of $F\mathfrak{A}^2$, so is not defined on all of $F\mathfrak{A}$.

In case (5), $U(\mathfrak{A}^2, N)$ is again the required neighbourhood. For here $p|F\mathfrak{A}^2$ maps $F\mathfrak{A}^2$ one-one onto $G\mathfrak{A}^2$ but is not a $\mathbb{C}^2$-morphism. Now $F\mathfrak{A}^2 \subseteq F\mathfrak{A}$ and $G\mathfrak{A}^2 \subseteq G\mathfrak{A}$, but no extension of $p|F\mathfrak{A}^2$ can be a $\mathbb{C}^2$-morphism from $F\mathfrak{A}$ onto $G\mathfrak{A}$, in particular $p|F\mathfrak{A}$ is not such a $\mathbb{C}^2$-morphism.

There remains case (3). We have $a \in F\mathfrak{A}^2$ and $p(a) \notin G\mathfrak{A}^2$. If for all $\mathfrak{A} \supseteq \mathfrak{A}^2$, $p(a) \notin G\mathfrak{A}$ then $U(\mathfrak{A}^2, N)$ is the required neighbourhood. Otherwise let $\mathfrak{A}^3 = \bigcap \{\mathfrak{A} \colon \mathfrak{A} \in U(\mathfrak{A}^2, N) \ \& \ p(a) \in G\mathfrak{A}\}$. $\mathfrak{A}^3$ is well-defined since we are taking the intersection of a set of objects in $\mathbb{C}^1$. Since $p(a) \in G\mathfrak{A}^3$, $G\mathfrak{A}^3 - G\mathfrak{A}^2 \neq \emptyset$ so $\mathfrak{A}^3 - \mathfrak{A}^2 \neq \emptyset$. Let $n \in \mathfrak{A}^3 - \mathfrak{A}^2$ and let $N^1 = N \cup \{n\}$. We claim $U(\mathfrak{A}^2, N^1)$ is the required neighbourhood. For if $\mathfrak{A} \in U(\mathfrak{A}^2, N^1)$, then $a \in F\mathfrak{A}$ but $p(a) \notin G\mathfrak{A}$, or else $p(a) \in G\mathfrak{A}$ and $\mathfrak{A}^3 \subseteq \mathfrak{A}$. But then $n \in \mathfrak{A}$ so $N^1 \cap \mathfrak{A} \neq \emptyset$ which is impossible. This completes the proof. $\square$

We shall need to "copy" neighbourhoods, so we prove the following corollary to the Duplication Lemma.

Lemma 18.3 (*Copying Lemma for neighbourhoods*). *Let* $U(\mathfrak{A}^0, M)$, $U(\mathfrak{B}^0, N)$ *be non-empty neighbourhoods, then there exist non-empty neighbourhoods*

$$U(\mathfrak{A}^1, M) \subseteq U(\mathfrak{A}^0, M)$$

$$U(\mathfrak{B}^1, N) \subseteq U(\mathfrak{B}^0, N)$$

and a one-one partial recursive function m such that for each $\mathfrak{B} \in U(\mathfrak{B}^1, N)$, m is defined on $\mathfrak{B}$, $m|\mathfrak{B}$ is a $\mathbb{C}$-isomorphism of $\mathfrak{B}$ onto some $\mathfrak{A} \in U(\mathfrak{A}^1, M)$.

Proof. By the Duplication Lemma 5.5 there exists $\mathfrak{B}^1 \supseteq \mathfrak{B}^0$ such that $\mathfrak{B}^1 \cap N = \emptyset$ where $\mathfrak{B}^1$ is $\mathbb{C}$-isomorphic to $\mathrm{cl}(\mathfrak{B}^0 \cup \mathfrak{A}^0)$. Again by the Duplication Lemma 5.5 we have an $\mathfrak{A}^1 \supseteq \mathfrak{A}^0$ with $\mathfrak{A}^1 \cap M = \emptyset$ and $\mathfrak{A}^1$ is $\mathbb{C}$-isomorphic to $\mathrm{cl}(\mathfrak{B}^0 \cup \mathfrak{A}^0)$. The Endomorphism Lemma 5.6 then yields a recursive monomorphism $m \colon \mathfrak{M} \to \mathfrak{M}$ such that m takes $\mathfrak{B}^1$ onto $\mathfrak{A}^1$ and $m(\mathfrak{M}) \cap M = \emptyset$. This m has the required properties. $\square$

Corollary 18.4. *With the same hypotheses as for Theorem 18.2, condition* (i) *is also equivalent to each of* (iv)–(vi) *below.*

(iv) *There is a finite* $Y \in \Omega(\mathbb{C}^1)$ *such that for all* $X \in \Omega(\mathbb{C}^1)$, $X \supseteq Y$ *implies* $F(X) = G(X)$,

(v) *For every finite* $Z \in \Omega(\mathbb{C}^1)$ *we can effectively find a finite* $Y \in \Omega(\mathbb{C}^1)$ *with* $Y \supseteq Z$ *such that for all* X, $X \supseteq Y$ *implies* $F(X) = G(X)$,

(vi) *For every non-empty neighbourhood* U *we can effectively find a non-empty subneighbourhood* U' *and a one-one partial recursive function* q *such that for all* $\mathfrak{A} \in U'$, q *is defined on* $F\mathfrak{A}$ *and* $q|F\mathfrak{A}$ *is a* $\mathbb{C}^2$*-morphism of* $F\mathfrak{A}$ *onto* $G\mathfrak{A}$.

Proof. Let X be a dense type, then for any finite Y we have $Y \subseteq X$. Hence if (iv) holds, (i) holds.

Trivially (v) implies (iv).

Now suppose (vi) holds; we shall prove (v). Suppose $Z \in \Omega(\mathbb{C}^1)$ and Z is finite. Choose $\mathfrak{Z} \in Z$ and let $U = U(\mathfrak{Z}, \emptyset)$. Let $U' = U(\mathfrak{Y}, N)$ and q be the non-empty subneighbourhood and partial recursive function given by (vi). Let $Y = \langle \mathfrak{Y} \rangle$ and suppose $X \supseteq Y$. Then there exist $\mathfrak{X}' \in X$ and $\mathfrak{Y}' \in Y$ with $\mathfrak{X}' \supseteq \mathfrak{Y}'$. Since $\mathfrak{Y}', N$ are finite by the Endomorphism Lemma 5.6, there exists $\mathfrak{X} \simeq \mathfrak{X}'$ such that $\mathfrak{X} \supseteq \mathfrak{Y}$ and $\mathfrak{X} \cap N = \emptyset$, that is $\mathfrak{X} \in U'$. But then if $q' = q|F\mathfrak{X}$, $q \colon F\mathfrak{X} \simeq G\mathfrak{X}$.

Finally we prove (iii) implies (vi). Let $U(\mathfrak{B}^0, N)$ be a non-empty neighbourhood on which F, G are uniformly effectively equivalent by p. Let $U(\mathfrak{A}^0, M)$ be an arbitrary non-empty neighbourhood. By the Copying Lemma 18.3 there is a recursive monomorphism m which maps each $\mathfrak{A}$ in a non-empty subneighbourhood $U(\mathfrak{A}^1, M)$ of $U(\mathfrak{A}^0, M)$ onto a $\mathfrak{B}$ in a non-empty subneighbourhood $U(\mathfrak{B}^1, N)$ of $U(\mathfrak{B}^0, N)$.

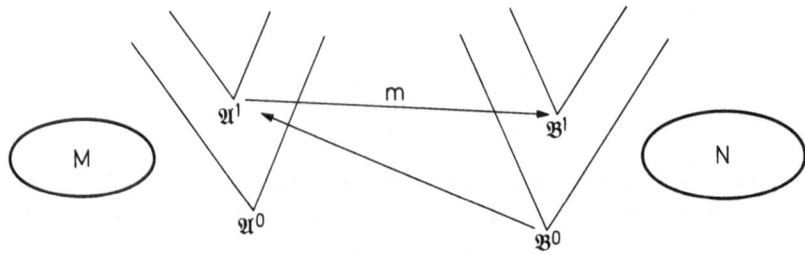

Let $\mathfrak{A} \in U(\mathfrak{A}^1, M)$, put $\mathfrak{B} = m\mathfrak{A}$. Let $m^{\mathfrak{A}}$ be the $\mathbb{C}^1$ isomorphism on $\mathfrak{A}$ to $\mathfrak{B}$ which is a restriction of m, let $p^{\mathfrak{A}}$ be the $\mathbb{C}^2$ isomorphism on $F(\mathfrak{B})$ to $G(\mathfrak{B})$ which is the restriction of p. Since F, G are combinatorial functors, both $Fm^{\mathfrak{A}}$ and $Gm^{\mathfrak{A}}$ are $\mathbb{C}^2$-isomorphisms. Define $q^{\mathfrak{A}} \colon F(\mathfrak{A}) \to G(\mathfrak{A})$ as

the isomorphism making

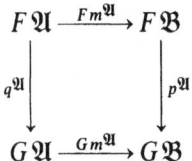

$$F\mathfrak{A} \xrightarrow{\;Fm^{\mathfrak{A}}\;} F\mathfrak{B}$$

commutative. Now suppose $\mathfrak{A}, \mathfrak{A}^1$ are both in $U(\mathfrak{A}^1, M)$. Then $m^{\mathfrak{A}}, m^{\mathfrak{A}^1}$ are both restrictions of m. So $m^{\mathfrak{A}} \cap m^{\mathfrak{A}^1}$ is defined, and by Theorem 3.12, $F(m^{\mathfrak{A}} \cap m^{\mathfrak{A}^1}) = F(m^{\mathfrak{A}}) \cap F(m^{\mathfrak{A}^1})$ and $G(m^{\mathfrak{A}} \cap m^{\mathfrak{A}^1}) = G(m^{\mathfrak{A}}) \cap G(m^{\mathfrak{A}^1})$. Thus $F(m^{\mathfrak{A}}), F(m^{\mathfrak{A}^1})$ are compatible and $G(m^{\mathfrak{A}}), G(m^{\mathfrak{A}^1})$ are compatible. The definition of $q^{\mathfrak{A}}, q^{\mathfrak{A}^1}$ now shows these are compatible too. We may therefore define a 1-1 function q as the union of graph $q^{\mathfrak{A}}$ for all finite $\mathfrak{A}$ in $U(\mathfrak{A}^1, M)$.

Suppose $\mathfrak{A}$ is finite, $\mathfrak{A} \in U(\mathfrak{A}^1, M)$. Since F, G are partial recursive, we find that $Fm^{\mathfrak{A}}, Gm^{\mathfrak{A}}$ are also partial recursive. So $F\mathfrak{A}, F\mathfrak{B}$ are recursively enumerable. Since p is partial recursive, we get $p^{\mathfrak{A}}$ partial recursive, uniformly in $\mathfrak{A}$. So q is partial recursive too. Since F, G are combinatorial functors, Theorem 2.5 and the definition of $q^{\mathfrak{A}}$ show that for any $\mathfrak{A}$ in $U(\mathfrak{A}^1, M)$, finite or not, $q^{\mathfrak{A}}$ is a restriction of q. So q has the desired properties for (vi). $\square$

Since finite isomorphism types coincide with finite RETs we immediately obtain:

Theorem 18.5. *Suppose every infinite type in $\Omega(\mathbb{C}^1)$ is dense (or that $\mathbb{C}^1$ has dimension). Then the following are equivalent when $F, G: \mathbb{C}^1 \to \mathbb{C}^2$ are partial recursive combinatorial functors.*

(i) *There is a finite* Y *such that for all* $X \in \Omega(\mathbb{C}^1)$, $X \supseteq Y$ *implies* $F(X) = G(X)$,

(ii) *There is a finite* Y *such that for all* $X \in \Lambda(\mathbb{C}^1)$, $X \supseteq Y$ *implies* $F(X) = G(X)$,

(iii) F, G *are uniformly effectively equivalent on a neighbourhood.* $\square$

Since partial recursive combinatorial functors are closed under composition (by Theorem 15.3) and F composed with G induces F composed with G (where F, G are the functions of RETs induced by F and G) it follows that we may replace F, G in the theorem by compositions of such functions. This is a generalization of Nerode [1961], Theorem 10.1 for the case of recursive combinatorial functions and series and also a partial generalization of Nerode [1966], Theorem 4.2. (For almost recursive combinatorial functions see §33.) Thus for the $\mathbb{S}$-RETs, that is, the RETs of Dekker, the equivalence of (i) and (ii) of Theorem 18.5 was known and the equivalence of (iii) was implicit in the proof. However, for other categories, in particular linear orderings, this result is new.

By the remarks following Lemma 17.4 the following categories satisfy the hypothesis of Theorem 18.5.

$\mathbb{S}$ the category of sets,
$\mathbb{L}$ the category of linear orderings,
$\mathbb{B}$ the category of Boolean algebras, and
$\mathbb{V}$ the category of vector spaces over a fixed finite field.

We end this section by strengthening Theorem 18.2 to a countable disjunction.

Corollary 18.6. *For $0 \leq i < \omega$ let F^{2i}, F^{2i+1} be partial recursive combinatorial functors from $\mathbb{C}$ to $\mathbb{C}^i$.*

If $\bigvee_i F^{2i}(X) = F^{2i+1}(X)$ is true for all Dedekind dense $\mathbb{C}$-types, then there exists i such that

$$F^{2i}(X) = F^{2i+1}(X)$$

is true for all Dedekind dense $\mathbb{C}$-types.

Proof. Suppose the conclusion is false. Then for each i there is a dense Dedekind $\mathbb{C}$-type X^i such that $F^{2i}(X^i) \neq F^{2i+1}(X^i)$. But then using the proof of Theorem 18.2, for each i, $\{\mathfrak{A} \in \mathbb{C} : F^{2i}\langle\mathfrak{A}\rangle = F^{2i+1}\langle\mathfrak{A}\rangle\}$ is nowhere dense. Hence

$$\{\mathfrak{A} \in \mathbb{C} : \exists i(F^{2i}\langle\mathfrak{A}\rangle = F^{2i+1}\langle\mathfrak{A}\rangle)\} \cup \{\mathfrak{A} \in \mathbb{C} : \mathfrak{A} \text{ is not Dedekind}\}$$

is a countable union of nowhere dense sets and its complement is therefore co-meagre, and in particular non-empty. That is, there exists $X \in \Lambda(\mathbb{C})$ such that for all i, $F^{2i}(X) \neq F^{2i+1}(X)$. $\square$

As we remarked after Theorem 18.5, we can replace the F^j be compositions of functions on $\Omega(\mathbb{C})$. We leave the reader to state the details formally.

19. More on Identities

In this section we illustrate some other techniques for treating identities on Dedekind types.

Corollary 19.1. *Suppose F, G are as in Theorem 18.2. Suppose $\mathfrak{M}^1$ is in a countable standard model M of Zermelo-Fraenkel set theory with the axiom of choice and the forcing conditions are the $U(\mathfrak{B}, N)$ neighbourhoods under inclusion. Let $\mathfrak{A}$ be a generic algebraically closed subsystem of $\mathfrak{M}^1$. Then each of (i)–(vi) is equivalent to*

(vii) $F(\langle\mathfrak{A}\rangle) = G(\langle\mathfrak{A}\rangle).$

Proof. The topological formulation of genericity says that generic sets are those not in any M-definable nowhere dense set. Also since F, G are partial recursive, $F(\mathfrak{X}) = \mathfrak{Y}$ are $G(\mathfrak{X}) = \mathfrak{Y}$ are absolute. ☐

Note. This illustrates how generic sets may be used to prove such results (*cf.* E. Ellentuck [1972]). Indeed any generic set gives a counter-example to *all* equations $F(X) = G(X)$ with $F(Y) \neq G(Y)$ for some dense type Y.

We earlier defined $A \subseteq B$ to hold if there exist $\mathfrak{A} \in A$ and $\mathfrak{B} \in B$ with $\mathfrak{A}$ a subsystem of $\mathfrak{B}$. This relation is trivially reflexive. If $\mathfrak{A} \subseteq \mathfrak{B}$, $p: \mathfrak{B} \simeq \mathfrak{B}'$ and $\mathfrak{B}' \subseteq \mathbb{C}'$, then $p(\mathfrak{A}) \subseteq \mathbb{C}'$ so $\subseteq$ is transitive on RETs. It is not antisymmetric on $\Omega(\mathbb{C})$ but does have this property on $\Lambda(\mathbb{C})$ as we now show. Note that it is trivially antisymmetric on finite RETs since any one-one map of a finite set into itself is onto. This is the fact that we use in the lemma below.

Example 19.1 (Dekker-Myhill [1960], p. 100). Let $\mathfrak{A}$ be a set which is not recursively enumerable but which contains an infinite recursively enumerable subset, $R = \mathbb{S}\text{-RET}(E)$ and $A = \mathbb{S}\text{-RET}(\mathfrak{A})$. Then trivially $\mathfrak{A} \subseteq E$ so $A \subseteq R$. But $\mathfrak{A}$ contains a subset which is the range of a one-one recursive (total) function so $E \simeq \mathfrak{B} \subseteq \mathfrak{A}$. Hence $R \subseteq A$. However, since $\mathfrak{A}$ is not recursively enumerable $R \neq A$.

Lemma 19.2. $\subseteq$ *is antisymmetric on* $\Lambda(\mathbb{C})$.

Proof. Suppose $p: \mathfrak{A} \simeq \mathfrak{B}' \subseteq \mathfrak{B}$ and $q: \mathfrak{B} \simeq \mathfrak{A}' \subseteq \mathfrak{A}$. Then $q \circ p$ maps $\mathfrak{A}$ into $\mathfrak{A}$. If $\mathfrak{A}$ is Dedekind then $q \circ p$, being one-one partial recursive, maps $\mathfrak{A}$ onto $\mathfrak{A}$. So p is onto and $\langle \mathfrak{A} \rangle = \langle \mathfrak{B} \rangle$. ☐

We shall now prove a variant of Theorem 18.2. Let $F, G: \mathbb{C}^1 \to \mathbb{C}^2$ be combinatorial functors then F is said to be *uniformly effectively embeddable* in G on a neighbourhood $U(\mathfrak{A}^1, N)$ if there is a one-one partial recursive function p such that for all $\mathfrak{A} \in U(\mathfrak{A}^1, N)$, p is defined on $F\mathfrak{A}$ and $p | F\mathfrak{A}$ is a $\mathbb{C}^2$-morphism of $F\mathfrak{A}$ into $G\mathfrak{A}$.

Theorem 19.3. *Let F, G satisfy the hypothesis of Theorem 18.2. Then the following conditions are equivalent:*

(i) $F(X) \subseteq G(X)$ *for all dense types X in* $\Omega(\mathbb{C}^1)$,

(ii) $F(X) \subseteq G(X)$ *for all Dedekind dense types in* $\Lambda(\mathbb{C}^1)$,

(iii) *F is uniformly effectively embeddable in G on some non-empty neighbourhood.*

Proof. We proceed as for Theorem 18.2. (iii) → (i) → (ii) are trivial as before. Suppose (iii) is false. Then for $U(\mathfrak{A}^1, N)$, any non-empty neighbourhood, and any one-one partial recursive p one of the following holds (with numbering as for Theorem 18.2) for some $\mathfrak{A}^2 \in U(\mathfrak{A}^1, N)$

(1) p is not defined on all of $F\mathfrak{A}^2$,

(3) for some $a \in F\mathfrak{A}^2$, $p(a) \notin G\mathfrak{A}^2$,

(5′) $p|F\mathfrak{A}^2$ maps $F\mathfrak{A}^2$ into $G\mathfrak{A}^2$ but is not a morphism of $\mathbb{C}^2$.

In case (1) $U(\mathfrak{A}^2, N)$ is the required subneighbourhood. In case (3) we repeat (3) of Theorem 18.2. In case (5′), $U(\mathfrak{A}^2, N)$ is the required neighbourhood for if $p|F\mathfrak{A}^2$ is not a $\mathbb{C}^2$-morphism then no extension of $p|F\mathfrak{A}^2$ can be.

It follows that $\{\mathfrak{A}: F\langle\mathfrak{A}\rangle \subseteq G\langle\mathfrak{A}\rangle\}$ is first category so (i) fails, and, as the non-Dedekind types are meagre, (ii) also fails.   □

Although, as we have seen $\subseteq$ is not antisymmetric on $\Omega(\mathbb{C})$ we do have the following corollary to Theorem 19.3.

Corollary 19.4. *Let F, G be as for the Theorem but be* finitary *recursive* (*so map* $^\circ\mathbb{C}^1$ *into* $^\circ\mathbb{C}^2$).

(i) *If* $\mathsf{F}(\mathsf{X}) \subseteq \mathsf{G}(\mathsf{X})$ *and* $\mathsf{G}(\mathsf{X}) \subseteq \mathsf{F}(\mathsf{X})$ *for all dense types* X *then* $\mathsf{F}(\mathsf{X}) = \mathsf{G}(\mathsf{X})$ *for all dense types* X.

(ii) *Similarly for all Dedekind dense types.*

Proof. By Theorem 19.3 (iii) there is a one-one partial recursive p and a non-empty neighbourhood $U(\mathfrak{A}^1, N^1)$ such that for all $\mathfrak{A} \in U(\mathfrak{A}^1, N^1)$, p is defined on $F(\mathfrak{A})$ and $p|F(\mathfrak{A})$ is a $\mathbb{C}^2$-morphism of $F(\mathfrak{A})$ into $G(\mathfrak{A})$. Similarly (by using the hypothesis $\mathsf{G}(\mathsf{X}) \subseteq \mathsf{F}(\mathsf{X})$) there is a non-empty neighbourhood $U(\mathfrak{A}^2, N^2)$ and a one-one partial recursive q such that for all $\mathfrak{A} \in U(\mathfrak{A}^2, N^2)$ q is defined on $G(\mathfrak{A})$ and $q|G(\mathfrak{A})$ is a $\mathbb{C}^2$-morphism of $G(\mathfrak{A})$ into $F(\mathfrak{A})$.

Now by the Copying Lemma 18.3 we may assume $U(\mathfrak{A}^2, N^2)$ is contained in a subneighbourhood of $U(\mathfrak{A}^1, N^1)$ then, for any $\mathfrak{A} \in U(\mathfrak{A}^2, N^2)$, $q \circ p$ is a $\mathbb{C}^2$-morphism of $F(\mathfrak{A})$ into $F(\mathfrak{A})$. But if $\mathfrak{A}$ is finite (as in the remark preceding Example 19.1) we must have that $q \circ p$ is onto. But F, G are continuous so, for all $\mathfrak{A} \in U(\mathfrak{A}^2, N^2)$, $q \circ p|F(\mathfrak{A})$ is a $\mathbb{C}^2$-morphism of $F(\mathfrak{A})$ onto $F(\mathfrak{A})$ and hence $F(\mathfrak{A}) \simeq G(\mathfrak{A})$ and the first part of the theorem follows. The second observation is immediate.   □

We strengthen part of Theorem 18.2 by using a more discriminating topological space but without requiring any conditions of density or dimension.

Theorem 19.5. *Let $\mathbb{C}^1$, $\mathbb{C}^2$ be suitable categories. Let F, $G: \mathbb{C}^1 \to \mathbb{C}^2$ be partial recursive combinatorial functors. If there exists an infinite $\mathsf{X} \in \Omega(\mathbb{C}^1)$ such that $\mathsf{F}(\mathsf{X}) \neq \mathsf{G}(\mathsf{X})$ then there are c Dedekind types Y such that $\mathsf{F}(\mathsf{Y}) \neq \mathsf{G}(\mathsf{Y})$ and the classical type of Y is that of X.*

Proof. Let $\mathfrak{A} \in X$ and let T be the set of all isomorphisms $q \in \mathbb{C}^1$ with $\operatorname{dom} q = A = |\mathfrak{A}|$. We take as basis for a topology $\mathcal{T}^1$ on T the sets

$$U(q^1, N^1) = \{q \in T : q \text{ extends } q^1 \ \& \ (\operatorname{codom} q) \cap N^1 = \emptyset\}$$

for q^1 a finite $\mathbb{C}^1$-isomorphism with $\operatorname{dom} q^1 \subseteq A$ and N^1 a finite set of natural numbers. (That the $U(q^1, N^1)$ form a basis follows easily as in the proof of Lemma 17.1.) Then T is a topological space to which Baire's category theorem applies.

Let Z be an infinite subset of $\mathfrak{M}^1$ then we claim

$$S_Z = \{q \in T : \operatorname{codom} q \supseteq Z\}$$

is nowhere dense in T. For let $U(q^1, N^1)$ be a non-empty basic neighbourhood. Since q^1 is finite and Z infinite there exists $a \in Z - \operatorname{codom} q^1$. Let $N^2 = N^1 \cup \{a\}$. Then $U(q^1, N^2)$ is a non-empty subneighbourhood contained in $U(q^1, N^1)$ not intersecting S_Z.

Say that, for p a one-one partial recursive function, p *matches* $\mathfrak{A}$ with $\mathfrak{B}$ if p restricted to $\mathfrak{A}$ is a $\mathbb{C}^2$-isomorphism of $\mathfrak{A}$ onto $\mathfrak{B}$. Then we claim that

$$S_p = \{q \in T : p \text{ matches } F(\operatorname{codom} q) \text{ with } G(\operatorname{dom} q)\}$$

is nowhere dense for any one-one partial recursive function p.

Let $U(q^1, N^1)$ be a non-empty basic neighbourhood. By the Endomorphism Lemma 5.6 there is a recursive $\mathbb{C}^1$-isomorphism q^2 extending q^1 such that $\operatorname{codom} q^2 \cap N^1 = \emptyset$. Since $A \subseteq |\mathfrak{M}^1|$, $X = \langle q^2(\mathfrak{A}) \rangle$ and by construction $q^2 \in U(q^1, N^1)$. By hypothesis $F(X) \neq G(X)$ so p does not match $F(q^2(\mathfrak{A}))$ with $G(q^2(\mathfrak{A}))$. If $\mathfrak{A}^1 = \operatorname{codom} q^1$, there exists a finite $\mathfrak{A}^2$ such that $\mathfrak{A}^1 \subseteq \mathfrak{A}^2 \subseteq q^2(\mathfrak{A})$ and p does not match $F(\mathfrak{A}^2)$ with $G(\mathfrak{A}^2)$.

Now we proceed word for word as in the proof of Theorem 18.2 from † (p. 62) where cases (1)–(5) may arise.

It follows that the set of $q \in T$ such that: either $\operatorname{codom} q$ contains an infinite recursively enumerable set or $F(\operatorname{codom} q)$ is matched with $G(\operatorname{codom} q)$ by some one-one partial recursive function p, is a countable union of nowhere dense sets. Hence it is meagre in T and its complement therefore has power c. That is, there exist $\mathbb{C}^1$-systems $\mathfrak{B}$ such that $F(\mathfrak{B}) \neq G(\mathfrak{B})$ but $\mathfrak{B}$ is $\mathbb{C}^1$-isomorphic to $\mathfrak{A}$. Since each $\mathbb{C}^1$-type contains only $\aleph_0$ objects of $\mathbb{C}^1$, taking $\mathbb{C}^1$-types completes the proof of the theorem. $\square$

Corollary 19.6. $F(X) = G(X)$ *for all* $X \in \Omega(\mathbb{C}^1)$ *if, and only if,* $F(X) = G(X)$ *for all* $X \in \Lambda(\mathbb{C}^1)$. *That is, identities between functions induced by partial recursive combinatorial functors are the same in* $\Omega(\mathbb{C}^1)$ *as in* $\Lambda(\mathbb{C}^1)$. $\square$

Corollary 19.7. *Let* $F, G : \mathbb{C}^1 \to \mathbb{C}^2$ *be partial recursive combinatorial functors. If for all* $X \in \Lambda(\mathbb{C}^1)$ *of a given classical type we have* $F(X) = G(X)$, *then for any Dedekind $\mathbb{C}$-type X there exists $\mathfrak{B}\in\mathbb{C}^1$ with $\mathfrak{B}\in X$ and a finite $\mathfrak{B}^1\subseteq\mathfrak{B}$ such that for all $\mathfrak{B}^2\in\mathbb{C}^1$, $\mathfrak{B}^1\subseteq\mathfrak{B}^2\subseteq\mathfrak{B}$ implies $F(\mathfrak{B}^2)\simeq G(\mathfrak{B}^2)$.*

Proof. The proof of Theorem 19.5 shows that there exists Y with the same classical type as X such that $F(Y)\neq G(Y)$ unless the following holds (*cf.* the part concerning S_p): There is a non-empty neighbourhood $U(q^1, N^1)$ in T, a one-one partial recursive function p and a one-one recursive function $q^2: \mathfrak{M}\to\mathbb{C}\in\mathbb{C}^1$ where $\mathbb{C}\in U(q^1, N^1)$ such that for all finite $\mathfrak{A}^2$ with $\mathfrak{A}^1\subseteq\mathfrak{A}^2\subseteq q^2(\mathfrak{A})$, p matches $F(\mathfrak{A}^2)$ with $G(\mathfrak{A}^2)$ where $\mathfrak{A}^1=\mathrm{codom}\,q^1$. Since F and G are continuous, p also matches $F(\mathfrak{A}^2)$ with $G(\mathfrak{A}^2)$, for all $\mathfrak{A}^2$ with $\mathfrak{A}^1\subseteq\mathfrak{A}^2\subseteq q^2(\mathfrak{A})$. Take $\mathfrak{B}^1=\mathfrak{A}^1$, $\mathfrak{B}^2$ for $\mathfrak{A}^2$ and $\mathfrak{B}=q^2(\mathfrak{A})$ then the proof is complete. $\square$

20. Uniform Implications for Dedekind Types

For $\mathfrak{M}$ a fully effective $\aleph_0$-categorical model let $\mathbb{E}(\mathfrak{M})$ be the set of all one-one partial functions with arguments and values in $\mathfrak{M}$. ($\mathbb{E}$ for embedding.) As we did for the objects of $\mathbb{C}$ we now put a (weak) topology on $\mathbb{E}$ by taking as basic neighbourhoods

$$U(p^0)=\{p\in\mathbb{E}: p\supseteq p^0\}$$

for p^0 a finite one-one map. If $L: \mathbb{E}\to\mathbb{E}$ is continuous then L is the (unique) continuous extension of $^\circ L=L$ restricted to $^\circ\mathbb{E}$, the set of finite one-one maps in $\mathbb{E}$ (*cf.* Theorem 2.5). A map $^\circ L: {}^\circ\mathbb{E}\to\mathbb{E}$ is continuous if, and only if, monotone – *i.e.* $p^1\subseteq p^2$ implies $^\circ L p^1\subseteq{}^\circ L p^2$.

Definition 20.1. $^\circ L: {}^\circ\mathbb{E}\to\mathbb{E}$ is said to be *partial recursive* if the function $^\lceil{}^\circ L^\rceil$ given by

$$^\lceil{}^\circ L^\rceil(^\lceil p^\rceil)= {}^\lceil{}^\circ L(p)^\rceil$$

is recursive where $^\lceil p^\rceil$ is an explicit index of (the finite map) p and $^\lceil{}^\circ L(p)^\rceil$ is a standard index of $^\circ L(p)$.

If $^\circ L: {}^\circ\mathbb{E}\to\mathbb{E}$ induces $L: \mathbb{E}\to\mathbb{E}$ by continuity we say that L is *partial recursive* if $^\circ L$ is.

We say that p *matches* $\mathfrak{A}$ with $\mathfrak{B}$ (by a $\mathbb{C}$-morphism) if $p|\mathfrak{A}$ maps $\mathfrak{A}$ onto $\mathfrak{B}$ (and $p|\mathfrak{A}$ is a $\mathbb{C}$-morphism).

In what follows we shall write $\mathbb{E}^i$ for $\mathbb{E}(\mathfrak{M}^i)$ where $\mathfrak{M}^i$ is the fully effective $\aleph_0$-categorical model giving rise to the category $\mathbb{C}^i$.

Definition 20.2. Let $F, G, H, K: \mathbb{C}^1\to\mathbb{C}^2$ be partial recursive combinatorial functors inducing functions F, G, H, K on RETs. Then an implication

$$F(X)=G(X)\to H(X)=K(X)$$

is said to be *uniformly effectively true for finite types* (by $°L$) if there is a continuous recursive operator

$$°L: \ °\mathbb{E}^2 \to \mathbb{E}^2$$

such that for all *finite* $\mathfrak{A} \in \mathbb{C}^1$ and all $p \in °\mathbb{E}^2$, if p matches $F(\mathfrak{A})$ with $G(\mathfrak{A})$ by a $\mathbb{C}^2$-morphism then $°L(p)$ matches $H(\mathfrak{A})$ with $K(\mathfrak{A})$ by a $\mathbb{C}^2$-morphism.

(We only use $\mathbb{C}^2$ for notational simplicity. We could have F, G: $\mathbb{C}^1 \to \mathbb{C}^2$ and H, K: $\mathbb{C}^1 \to \mathbb{C}^3$ with corresponding changes for the $\mathbb{E}^i$. Further we could use

$$F^1(X) = G^1(X) \ \& \cdots \& \ F^n(X) = G^n(X) \to H(X) = K(X):$$

we should only need to regard p as an n-tuple. Later on we shall discuss the case of an infinite conjunction.)

Theorem 20.3. *Suppose*

(1) $$F(X) = G(X) \to H(X) = K(X)$$

is uniformly effectively true for finite types then

$$F(X) = G(X) \to H(X) = K(X)$$

is true for all Dedekind types.

Proof. Let (1) be true by $°L: \ °\mathbb{E}^2 \to \mathbb{E}^2$ and let L be the continuous extension of $°L$ to $\mathbb{E}^2$.

Let $\mathfrak{A}$ be Dedekind. We shall show that a match of $F(\mathfrak{A})$ with $G(\mathfrak{A})$ implies a match of $F(\mathfrak{B})$ with $G(\mathfrak{B})$ for enough finite $\mathfrak{B}$ so that by applying $°L$ to such we can piece together a matching of $H(\mathfrak{A})$ with $K(\mathfrak{A})$.

Lemma 20.4 (Assuming $\mathfrak{A}$ is Dedekind). *Let A^0 be a finite subset of $\mathfrak{A}$ and suppose p matches $F(\mathfrak{A})$ with $G(\mathfrak{A})$ by a $\mathbb{C}^2$-morphism. Then there is a finite object $\mathfrak{B}$ such that $A^0 \subseteq \mathfrak{B} \subseteq \mathfrak{A}$ such that p matches $F(\mathfrak{B})$ with $G(\mathfrak{B})$ by a $\mathbb{C}^2$-morphism.*

Proof. We give an effective inductive definition so that B (which will be the universe of $\mathfrak{B}$) is the smallest set satisfying 1.–4. below.

1. $A^0 \subseteq B$,
2. $\mathrm{cl}(B) \subseteq B$,
3. If $F^{\leftarrow}(b) \subseteq B$ then $G^{\leftarrow}(p(b)) \subseteq B$,
4. If $G^{\leftarrow}(b) \subseteq B$ then $F^{\leftarrow}(p^{-1}(b)) \subseteq B$.

All clauses are effective since $F^{\leftarrow}(b)$ and $G^{\leftarrow}(b)$ are finite sets effectively computable from b (*cf.* Section 15). Hence B is recursively enumerable. But $\mathfrak{A}$ is Dedekind and $B \subseteq \mathfrak{A}$ so B is finite. B is algebraically closed by 2.

Let $\mathfrak{B}$ be the substructure of $\mathfrak{A}$ with universe B. We now show that 3 and 4 guarantee that p matches $F(\mathfrak{B})$ with $G(\mathfrak{B})$. Suppose $b \in F(\mathfrak{B})$ then $F^-(b) \subseteq \mathfrak{B}$ so by 3, $G^-(p(b)) \subseteq B$. But then $p(b) \in G(\mathfrak{B})$. Similarly if $b \in G(\mathfrak{B})$ then $G^-(b) \subseteq \mathfrak{B}$ so by 4, $F^-(p^{-1}(b)) \subseteq \mathfrak{B}$. But then $p^{-1}(b) \in F(\mathfrak{B})$.

Finally, $p|F(\mathfrak{B})$ is a restriction to an algebraically closed set of a $\mathbb{C}^2$-morphism since p matches $F(\mathfrak{A})$ with $G(\mathfrak{A})$ by a $\mathbb{C}^2$-morphism. Hence $p|F(\mathfrak{B})$ is a $\mathbb{C}^2$-morphism and the lemma is established. $\square$

We now complete the proof of the theorem.

Suppose $a \in H(\mathfrak{A})$, then, since H is continuous $a \in H(\mathfrak{A}^0)$ for some finite $\mathfrak{A}^0 \subseteq \mathfrak{A}$. By the lemma there is a finite $\mathfrak{B} \supseteq \mathfrak{A}^0$ such that $\mathfrak{B} \subseteq \mathfrak{A}$ and p matches $F(\mathfrak{B})$ with $G(\mathfrak{B})$ by a $\mathbb{C}^2$-morphism. Call this morphism p^0 then, since L is continuous, $L(p)(a) = L(p^0)(a) \in K(\mathfrak{B}) \subseteq K(\mathfrak{A})$. So $L(p)$ maps $H(\mathfrak{A})$ into $K(\mathfrak{A})$. Similarly, $(L(p))^{-1}$ maps $K(\mathfrak{A})$ into $H(\mathfrak{A})$ and therefore $L(p)|H(\mathfrak{A})$ maps $H(\mathfrak{A})$ onto $K(\mathfrak{A})$. Finally we show $L(p)|H(\mathfrak{A})$ is a $\mathbb{C}^2$-morphism. Since all relations are finitary it suffices to show that if C^0 is any finite subset of $H(\mathfrak{A})$ then $L(p)|\mathrm{cl}\,C^0$ is a $\mathbb{C}^2$-morphism. So obtain a finite $B \supseteq C^0$ from the lemma with $|\mathfrak{B}| = B$ and then $p' = p|F(\mathfrak{B})$ is a $\mathbb{C}^2$-morphism from $F(\mathfrak{B})$ to $G(\mathfrak{B})$ so $L(p')|H(\mathfrak{B})$ is a $\mathbb{C}^2$-morphism from $H(\mathfrak{B})$ to $K(\mathfrak{B})$. But L is continuous so $L(p') \subseteq L(p)$. We therefore conclude that $L(p)|H(\mathfrak{A})$ is indeed a $\mathbb{C}^2$-morphism from $H(\mathfrak{A})$ onto $K(\mathfrak{A})$. $\square$

Now say that an implication

$$\mathsf{F}(\mathsf{X}) \subseteq \mathsf{G}(\mathsf{X}) \to \mathsf{H}(\mathsf{X}) \subseteq \mathsf{K}(\mathsf{X})$$

is *uniformly effectively true for finite types* (by $^\circ L$) if there is a continuous recursive operator

$$^\circ L: \; ^\circ \mathbb{E}^2 \to \mathbb{E}^2$$

such that for all finite $\mathfrak{A} \in \mathbb{C}^1$ and all $p \in {}^\circ\mathbb{E}^2$ if p matches $F(\mathfrak{A})$ with some $\mathfrak{B} \subseteq G(\mathfrak{A})$ by a $\mathbb{C}^2$-morphism then $^\circ L(p)$ matches $H(\mathfrak{A})$ with a subsystem of $K(\mathfrak{A})$ by a $\mathbb{C}^2$-morphism.

Corollary 20.5. *Suppose*

$$\mathsf{F}(\mathsf{X}) \subseteq \mathsf{G}(\mathsf{X}) \to \mathsf{H}(\mathsf{X}) \subseteq \mathsf{K}(\mathsf{X})$$

is uniformly effectively true for finite types, then

$$\mathsf{F}(\mathsf{X}) \subseteq \mathsf{G}(\mathsf{X}) \to \mathsf{H}(\mathsf{X}) \subseteq \mathsf{K}(\mathsf{X})$$

is true for all Dedekind types.

Proof. We proceed as in the proof of Theorem 20.3 but dropping clause 4 of the inductive definition.

We then only need to change "matches $F(\mathfrak{B})$ with $G(\mathfrak{B})$" to "matches $F(\mathfrak{B})$ with a subsystem of $G(\mathfrak{B})$" and similarly for H and K.

Dropping 4 only allows us to infer that $L(p)$ maps $H(\mathfrak{A})$ into $K(\mathfrak{A})$. □

We leave as an exercise the task of formulating and proving the results for implications involving both $\subseteq$ and $=$. We merely remark that one special case is the antisymmetry of $\subseteq$, that is to say

$$\forall X (X \subseteq Y \ \& \ Y \subseteq X \rightarrow X = Y).$$

For here we need only take $F(X, Y) = X$ and $G(X, Y) = Y$ and the corresponding assertion is always true for finite types so it is true for Dedekind types (in any suitable category).

Part VI. Frames

21. Frames

The method of frames originated in Nerode [1961]. There frames were used to extend recursively enumerable relations on the natural numbers to Dedekind $\mathbb{S}$-RETs (isols). The main effect of the method is to obtain effective properties of Dedekind $\mathbb{S}$-RETs by approximating them effectively by using finite sets. The main application is the reduction of questions about Dedekind $\mathbb{S}$-RETs to questions about natural numbers. The method generalizes to all the categories we have considered but the results are most striking (and most obviously the generalizations of earlier results) in the case where the categories involved have dimension. However, in many other cases a weaker property (the automorphism extension property) allows us to obtain full analogues of Nerode's [1961] results. But first we need the machinery.

Since a relation of the form $\{(\mathfrak{A}^0, \ldots, \mathfrak{A}^n): \mathfrak{A}^i \in \mathbb{C}^i$ for $i < n$ and $(\mathfrak{A}^0, \ldots, \mathfrak{A}^n) \in R\}$ may be identified with a subset of the product category $\mathbb{C}^0 \times \cdots \times \mathbb{C}^n$ we shall in general treat sets rather than relations. As usual we write $A \subseteq \mathfrak{B}$ instead of $A \subseteq |\mathfrak{B}|$.

Definition 21.1. A *frame* is a subset F of $\mathbb{C}$ such that

(i) $\mathfrak{A} \in F$ implies $\mathfrak{A}$ is finite,

(ii) $\mathfrak{A}, \mathfrak{B} \in F$ implies $\mathfrak{A} \cap \mathfrak{B} \in F$.

An object $\mathfrak{A} \in \mathbb{C}$ is said to be *attainable* from a frame F if, for any finite set $B \subseteq \mathfrak{A}$, there exists $\mathfrak{C} \in F$ with $B \subseteq \mathfrak{C} \subseteq \mathfrak{A}$. We write $\mathscr{A}(F)$ for the collection of all $\mathfrak{A} \in \mathbb{C}$ which are attainable from the frame F.

We observe that if $\mathfrak{B}$ is finite and attainable from a frame F then $\mathfrak{B} \in F$. We also note that $\subseteq$ is a partial order on $\mathscr{A}(F)$ and that greatest lower bounds (g.l.b.) and least upper bounds (l.u.b.) sometimes exist. We write g.l.b. S and l.u.b. S for the g.l.b. and l.u.b. (respectively) of a set $S \subseteq F$.

Lemma 21.2. *Let* F *be a frame.*

(i) *If* $\mathfrak{B}$ *is attainable from* F *and* $A \subseteq \mathfrak{B}$ *then* A *has a* l.u.b. *in* $\mathscr{A}(F)$.

(ii) *If each* $\mathfrak{A}^i$ *(for* $i \in I$*) is attainable from* F *and* $I \neq \emptyset$ *then* g.l.b. $\{\mathfrak{A}^i : i \in I\} = \bigcap \{\mathfrak{A}^i : i \in I\}$ *and is attainable from* F.

Proof. If $\mathfrak{B} \in \mathscr{A}(F)$ then for each finite $D \subseteq \mathfrak{B}$ there is a minimum $\mathfrak{C}$ such that $\mathfrak{C} \in F$ and $D \subseteq \mathfrak{C} \subseteq \mathfrak{B}$. (Take a finite $\mathfrak{C}^0$ with $D \subseteq \mathfrak{C}^0 \subseteq \mathfrak{B}$ which exists since $\mathfrak{B} \in \mathscr{A}(F)$ and intersect it with all other $\mathfrak{C}$ such that $D \subseteq \mathfrak{C} \subseteq \mathfrak{B}$ and $\mathfrak{C} \in F$: though in fact only a finite intersection is required as $\mathfrak{C}^0$ is finite.)

Let $\mathfrak{A}^0$ be the union of all the minimum $\mathfrak{C}$ such that $\mathfrak{C} \in F$ and $D \subseteq \mathfrak{C} \subseteq \mathfrak{B}$ for some finite $D \subseteq A$. Clearly $A \subseteq \mathfrak{A}^0$. By construction $\mathfrak{A}^0 \in \mathscr{A}(F)$, and is the least such member of $\mathscr{A}(F)$. Thus (i) is established.

(ii) Let B be finite and $B \subseteq \bigcap \{\mathfrak{A}^i : i \in I\}$ where $I \neq \emptyset$ and each $\mathfrak{A}^i \in \mathscr{A}(F)$. Then for each i there exists $\mathfrak{C}^i \in F$ such that $B \subseteq \mathfrak{C}^i \subseteq \mathfrak{A}^i$. Let $\mathfrak{C}^0$ be one such $\mathfrak{C}^i$ (which exists since $I \neq \emptyset$), then there are only a finite number of distinct $\mathfrak{D}^j = \mathfrak{C}^0 \cap \mathfrak{C}^j$ since there are only a finite number of (finite) D with $D \subseteq \mathfrak{C}^0$. Hence $\mathfrak{D}^I = \bigcap \{\mathfrak{C}^i : i \in I\}$ is finite and in F being a finite intersection of elements in F. Clearly $\mathfrak{D}^I \subseteq \bigcap \{\mathfrak{A}^i : i \in I\}$. But $B \subseteq \mathfrak{D}^I$ and (ii) is proved. $\square$

We write $C_F(A)$ for the least upper bound in F of A if A has some upper bound in F. $C_F(A)$ is called the F-*closure* of A (or closure of A in F). We write $C_F(\mathfrak{A})$ for $C_F(|\mathfrak{A}|)$.

We write
$$F^* = \{B : (\exists \mathfrak{A} \in F)(B \subseteq \mathfrak{A})\}.$$

So F* consists of all those sets which can be extended to (universes of) members of F and $B \in F^*$ implies $C_F(B)$ is defined.

As usual we assume we have a fully effective explicit listing of finite subsets of E (which is the universe of $\mathfrak{M}$). As before $\ulcorner A \urcorner$ denotes the explicit index of A; similarly for $\ulcorner \mathfrak{A} \urcorner$.

Definition 21.3. A frame F is said to be *recursive* if

(i) the set $\ulcorner F^* \urcorner = \{\ulcorner A \urcorner : A \in F^*\}$ is recursively enumerable and

(ii) the function $\ulcorner C_F \urcorner$ whose domain is $\ulcorner F^* \urcorner$ is partial recursive where
$$\ulcorner C_F \urcorner (\ulcorner A \urcorner) = \ulcorner C_F(A) \urcorner.$$

(N.B. $\ulcorner C_F(A) \urcorner$ is an *explicit* index of the finite system $C_F(A)$.)

Corollary 21.4. *If* F *is a recursive frame then*
$$\ulcorner F \urcorner = \{\ulcorner \mathfrak{A} \urcorner : \mathfrak{A} \in F\}$$
is recursively enumerable.

Proof. $\ulcorner F \urcorner = $ range of $\ulcorner C_F \urcorner = \ulcorner C_F \urcorner (\ulcorner F^* \urcorner)$. $\square$

Example 21.1. $^\circ \mathfrak{C}$ is a recursive frame. F* is the set of all finite subsets of $|\mathfrak{M}|$. $C_F(A)$ is the algebraic closure of A. $\mathscr{A}(F) = \mathfrak{C}$.

Example 21.2. $\emptyset$ is a recursive frame. $\mathscr{A}(\emptyset)=\emptyset$. $\{\mathfrak{A}\}$ is a recursive frame if $\mathfrak{A}\in\mathbb{C}$. $\mathscr{A}(\mathfrak{A})=\{\mathfrak{A}\}$.

Example 21.3. If F is a recursive frame and $\mathfrak{A}\in F$ then $\{\mathfrak{B}\in F: \mathfrak{A}\subseteq\mathfrak{B}\}$ is a recursive frame.

Theorem 21.5. *Let p be a one-one partial recursive function such that* $p: \mathfrak{A}\simeq\mathfrak{B}$. *Then* $F=\{\mathbb{C}\in{}^\circ\mathbb{C}: \mathbb{C}\subseteq\mathrm{dom}\,p$ *and* $p|\mathbb{C}$ *is a* $\mathbb{C}$*-isomorphism*$\}$ *is a recursive frame from which* $\mathfrak{A}$ *is attainable.*

Proof. F contains with any object also any smaller object, so F is a frame. A finite set C^0 of natural numbers is in F* if and only if $\mathrm{cl}(C^0)$ exists and $\mathrm{cl}(C^0)\subseteq\mathrm{dom}\,p$ and $p|\mathrm{cl}(C^0)$ is a $\mathbb{C}$-isomorphism. Then $C_F(C^0)$ is $\mathrm{cl}(C^0)$. Note that $\ulcorner\mathrm{cl}(C^0)\urcorner$ can be computed from $\ulcorner C^0\urcorner$, also note that $\mathrm{cl}(C^0)\subseteq\mathrm{dom}\,p$ can eventually be verified if true, and finally then whether $p|\mathrm{cl}(C^0)$ is a $\mathbb{C}$-isomorphism can be checked. So F* is recursively enumerable. Finally we show $\mathfrak{A}\in\mathscr{A}(F)$. If C^0 is a finite subset of $|\mathfrak{A}|$, we must find a $B\in F$ with $C^0\subseteq B\subseteq|\mathfrak{A}|$. But $B=\mathrm{cl}(C^0)$ will do. $\square$

22. Frame Maps are Map Frames

In this section we establish in essence that a certain kind of map, namely a frame map, is a map which is attainable from a frame.

Definition 22.1. Let $\mathbb{D}\subseteq\mathbb{C}^1$ and let $\Phi: \mathbb{D}\to\mathbb{C}^2$ then Φ is said to be a *frame map* if

(i) $\mathbb{D}=\mathscr{A}(F)$ for some frame $F\subseteq{}^\circ\mathbb{C}^1$,

(ii) Φ is finitary, that is, if $\mathfrak{A}$ is finite, $\Phi(\mathfrak{A})$ is finite (if defined) and

(iii) if $x\in\Phi(\mathfrak{A})$ for some $\mathfrak{A}\in\mathbb{D}$ then there is a finite object $\Phi^-(x)$ in $\mathbb{D}$, called the *pseudo-inverse* of x, such that for all $\mathfrak{B}\in\mathbb{D}$,

$$x\in\Phi(\mathfrak{B}) \quad \text{if, and only if, } \Phi^-(x)\subseteq\mathfrak{B}.$$

We now proceed to show that a map $\Phi: \mathbb{D}\to\mathbb{C}^2$ is a frame map if, and only if, the graph of Φ is $\mathscr{A}(G)$ for some frame G. But before we do this we note the following

Corollary 22.2. *A finitary combinatorial functor* $F: \mathbb{C}^1\to\mathbb{C}^2$ *is a frame map.*

Proof. Domain of $F=\mathscr{A}({}^\circ\mathbb{C}^1)$ by Example 21.2, so (i) is satisfied. (ii) is immediate as F is finitary. For (iii) take $F^-(x)$ as in the proof of Theorem 3.12, that is $F^-(x)=\bigcap\{\mathfrak{A}\in{}^\circ\mathbb{C}^1: x\in F(\mathfrak{A})\}$. $\square$

Lemma 22.3. *Let Φ be a frame map. Then Φ preserves intersections of objects and is continuous.*

Proof. $x \in \Phi(\bigcap\{\mathfrak{A}^i : i \in I\})$

if and only if $\Phi^\leftarrow(x) \subseteq \bigcap\{\mathfrak{A}^i : i \in I\}$ by 22.1 (iii)

if and only if $\Phi^\leftarrow(x) \subseteq \mathfrak{A}^i$ for all $i \in I$

if and only if $x \in \Phi(\mathfrak{A}^i)$ for all $i \in I$

if and only if $x \in \bigcap\{\Phi(\mathfrak{A}^i) : i \in I\}$.

Hence Φ preserves intersections.

By condition 22.1 (iii) if $x \in \Phi(\mathfrak{B})$ then $x \in \Phi(\Phi^\leftarrow(x))$ and $\Phi^\leftarrow(x)$ is finite, so for all $\mathfrak{A}$ in dom Φ

$$\Phi(\mathfrak{A}) \subseteq \bigcup\{\Phi(\mathfrak{B}) : \mathfrak{B} \subseteq \mathfrak{A} \ \& \ \mathfrak{B} \in F\}.$$

Since Φ preserves intersections

$$\Phi(\mathfrak{A} \cap \mathfrak{B}) = \Phi(\mathfrak{A}) \cap \Phi(\mathfrak{B}).$$

So $\mathfrak{B} \subseteq \mathfrak{A}$ implies $\Phi(\mathfrak{B}) \subseteq \Phi(\mathfrak{A})$. Hence we have

$$\Phi(\mathfrak{A}) = \bigcup\{\Phi(\mathfrak{B}) : \mathfrak{B} \subseteq \mathfrak{A} \ \& \ \mathfrak{B} \in F\}$$

and therefore Φ is continuous. $\square$

Corollary 22.4. *If Φ is a frame map then $\mathfrak{A} \subseteq \mathfrak{B}$ implies $\Phi(\mathfrak{A}) \subseteq \Phi(\mathfrak{B})$.* $\square$

Consider the graph of Φ. If $\Phi: \mathbb{D} \to \mathbb{C}^2$ where $\mathbb{D} \subseteq \mathbb{C}^1$ then graph of $\Phi \subseteq \mathbb{C}^1 \times \mathbb{C}^2$. (We shall identify Φ with its graph where convenient and therefore we may write $\Phi \subseteq \mathbb{C}^1 \times \mathbb{C}^2$.)

Theorem 22.5. *Suppose $\mathbb{D} \subseteq \mathbb{C}^1$ and $\Phi: \mathbb{D} \to \mathbb{C}^2$. Then Φ is a frame map if, and only if, there is a frame $G \subseteq {}^\circ\mathbb{C}^1 \times {}^\circ\mathbb{C}^2$ such that $\Phi = \mathscr{A}(G)$.*

Proof. Suppose Φ is a frame map and dom $\Phi = \mathbb{D} = \mathscr{A}(F)$ for some frame $F \subseteq {}^\circ\mathbb{C}^1$. Let $G = \Phi \cap (F \times {}^\circ\mathbb{C}^2)$. We claim that G is a frame and $\Phi = \mathscr{A}(G)$.

By construction $G \subseteq {}^\circ\mathbb{C}^1 \times {}^\circ\mathbb{C}^2 = {}^\circ(\mathbb{C}^1 \times \mathbb{C}^2)$. The elements of G are of the form $(\mathfrak{A}, \Phi(\mathfrak{A}))$ with $\mathfrak{A} \in {}^\circ\mathbb{C}^1$ and $\Phi(\mathfrak{A}) \in {}^\circ\mathbb{C}^2$. Therefore to show G is a frame we only have to show G is closed under intersection.

Now $\bigcap\{(\mathfrak{A}^i, \Phi(\mathfrak{A}^i)) : i \in I\} = (\bigcap \mathfrak{A}^i, \bigcap \Phi(\mathfrak{A}^i)) = (\bigcap \mathfrak{A}^i, \Phi(\bigcap \mathfrak{A}^i))$ by Lemma 22.3 so G is indeed closed under intersections.

We now prove the two inclusions (a) $\Phi \subseteq \mathscr{A}(G)$ and (b) $\mathscr{A}(G) \subseteq \Phi$. $\mathbb{C}^i$ arises from a fully effective $\aleph_0$-categorical model $\mathfrak{M}^i$ so to ease notation we write $(C, D) \in {}^\circ(\mathbb{C}^1 \times \mathbb{C}^2)$ if $C \subseteq |\mathfrak{M}^1|$ and $D \subseteq |\mathfrak{M}^2|$, and similarly $(C, D) \subseteq (\mathfrak{A}, \mathfrak{B})$ if $C \subseteq |\mathfrak{A}|$ and $D \subseteq |\mathfrak{B}|$, where $\mathfrak{A} \subseteq \mathfrak{M}^1$ and $\mathfrak{B} \subseteq \mathfrak{M}^2$.

(a) Suppose $(C, D) \subseteq (\mathfrak{A}, \Phi(\mathfrak{A}))$ where $(C, D) \in {}^\circ(\mathbb{C}^1 \times \mathbb{C}^2)$. We need to show there exists $\mathfrak{E} \in F$ with $\mathfrak{E} \subseteq \mathfrak{A}$ and $(C, D) \subseteq (\mathfrak{E}, \Phi(\mathfrak{E}))$.

$D \subseteq \Phi(\mathfrak{A})$ implies $\Phi^{\leftarrow}(x) \subseteq \mathfrak{A}$ for all $x \in D$ by 22.1 (iii) and each $\Phi^{\leftarrow}(x)$ is finite.

Let $\mathfrak{E} = C_F\big(C \cup \bigcup\{\Phi^{\leftarrow}(x): x \in D\}\big)$.

Since C and all the $\Phi^{\leftarrow}(x)$ are finite and contained in $\mathfrak{A}$ it follows that $C \cup \bigcup\{\Phi^{\leftarrow}(x): x \in D\}$ is finite and contained in $\mathfrak{A}$. Now $\mathfrak{A} \in \mathscr{A}(F)$ so $\mathfrak{E}$ exists and $\mathfrak{E} \subseteq \mathfrak{A}$. But we also have $D \subseteq \Phi(\mathfrak{E})$ since for each $x \in D$, $\Phi^{\leftarrow}(x) \subseteq \mathfrak{E}$. Corollary 22.4 gives $\Phi(\mathfrak{E}) \subseteq \Phi(\mathfrak{A})$ and hence

$$(C, D) \subseteq (\mathfrak{E}, \Phi(\mathfrak{E})) \subseteq (\mathfrak{A}, \Phi(\mathfrak{A}))$$

and therefore

$$(\mathfrak{A}, \Phi(\mathfrak{A})) \in \mathscr{A}(G).$$

(b) Now suppose $(\mathfrak{A}, \mathfrak{B}) \in \mathscr{A}(G)$. We wish to show $\mathfrak{A} \in \mathbb{D} = \mathscr{A}(F)$ and $\mathfrak{B} = \Phi(\mathfrak{A})$.

Let C be finite and $C \subseteq \mathfrak{A}$. Then $(C, \emptyset) \subseteq (\mathfrak{A}, \mathfrak{B})$. Since $(\mathfrak{A}, \mathfrak{B})$ is attainable from G there exists a finite $\mathfrak{E} \in F$ such that

$$(C, \emptyset) \subseteq (\mathfrak{E}, \Phi(\mathfrak{E})) \subseteq (\mathfrak{A}, \mathfrak{B}).$$

But then $C \subseteq \mathfrak{E} \subseteq \mathfrak{A}$ so $\mathfrak{A} \in \mathscr{A}(F)$.

Since $(\mathfrak{A}, \mathfrak{B}) \in \mathscr{A}(G)$,

$$(\mathfrak{A}, \mathfrak{B}) = \bigcup\{(\mathfrak{C}, \Phi(\mathfrak{C})): \mathfrak{C} \subseteq \mathfrak{A} \ \& \ \mathfrak{C} \in F\}.$$

But Φ is a frame map and therefore by Lemma 22.3 is continuous so $\bigcup\{\Phi(\mathfrak{C}): \mathfrak{C} \subseteq \mathfrak{A} \ \& \ \mathfrak{C} \in F\} = \Phi(\mathfrak{A})$. So $\mathfrak{B} = \Phi(\mathfrak{A})$ as required.

Now we prove the other half of the theorem, that is, that if $\Phi = \mathscr{A}(G)$ then Φ is a frame map.

We check the conditions of Definition 22.1 in the order (ii), (i), (iii).

(ii) Suppose $(\mathfrak{A}, \mathfrak{B}) \in \Phi$ and $\mathfrak{A}$ is finite. Then $(\mathfrak{A}, \emptyset)$ is finite and $(\mathfrak{A}, \emptyset) \subseteq (\mathfrak{A}, \mathfrak{B})$. Since $(\mathfrak{A}, \mathfrak{B}) \in \mathscr{A}(G) = \Phi$ there exists a finite $(\mathfrak{C}, \mathfrak{D}) \in G$ such that

$$(\mathfrak{A}, \emptyset) \subseteq (\mathfrak{C}, \mathfrak{D}) \subseteq (\mathfrak{A}, \mathfrak{B}).$$

It follows that $\mathfrak{A} = \mathfrak{C}$. But Φ, by hypothesis, is a map so is single valued and therefore

$$\mathfrak{B} = \Phi(\mathfrak{A}) = \mathfrak{D}.$$

But then $\mathfrak{B}$ is finite, so Φ is finitary.

(i) Set $F = \mathbb{D} \cap {}^{\circ}\mathbb{C}^1$. Trivially F contains only finite elements. Suppose $\mathfrak{C}, \mathfrak{D} \in F$ then $(\mathfrak{C}, \Phi(\mathfrak{C}))$ and $(\mathfrak{D}, \Phi(\mathfrak{D}))$ are in $\mathscr{A}(G)$ and are finite by (ii) so they are both in G.

Now $$(\mathfrak{C}, \Phi(\mathfrak{C})) \cap (\mathfrak{D}, \Phi(\mathfrak{D})) = (\mathfrak{C} \cap \mathfrak{D}, \Phi(\mathfrak{C}) \cap \Phi(\mathfrak{D})).$$

But G, being a frame, is closed under intersection. So

$$(\mathfrak{C} \cap \mathfrak{D}, \Phi(\mathfrak{C}) \cap \Phi(\mathfrak{D})) \in G$$

and $\mathfrak{C} \cap \mathfrak{D} \in F$. Therefore, F is a frame.

Next we show that $\mathbb{D}=\mathscr{A}(F)$. First of all, $\mathbb{D}\subseteq\mathscr{A}(F)$. For suppose $\mathfrak{A}\in\mathbb{D}$ and C is finite with $C\subseteq\mathfrak{A}$. Then $(C,\emptyset)\subseteq(\mathfrak{A},\Phi(\mathfrak{A}))$. Therefore there exists $\mathfrak{D}\in{}^\circ\mathbb{C}^1$ such that

$$(C,\emptyset)\subseteq(\mathfrak{D},\Phi(\mathfrak{D}))\subseteq(\mathfrak{A},\Phi(\mathfrak{A})).$$

Hence $C\subseteq\mathfrak{D}\subseteq\mathfrak{A}$ so $\mathfrak{A}\in\mathscr{A}(F)$.

Conversely, suppose $\mathfrak{A}\in\mathscr{A}(F)$. Then

$$\mathfrak{A}=\bigcup\{\mathfrak{C}:\mathfrak{C}\subseteq\mathfrak{A}\ \&\ \mathfrak{C}\in F\}.$$

Set

$$\mathfrak{B}=\bigcup\{\Phi(\mathfrak{C}):\mathfrak{C}\subseteq\mathfrak{A}\ \&\ \mathfrak{C}\in F\}.$$

We now have to show $(\mathfrak{A},\mathfrak{B})\in\mathscr{A}(G)$ for then we shall have $\mathfrak{B}=\Phi(\mathfrak{A})$ so $\mathfrak{A}\in\mathscr{A}(F)$.

Suppose $(C,D)\subseteq(\mathfrak{A},\mathfrak{B})$ and (C,D) is finite. Let $x\in D$ then, because $x\in\mathfrak{B}$ there exists $\mathfrak{C}_x\in F$ such that $x\in\Phi(\mathfrak{C}_x)$ and

$$(\emptyset,\{x\})\subseteq(\mathfrak{C}_x,\Phi(\mathfrak{C}_x))\subseteq(\mathfrak{A},\mathfrak{B}).$$

Let $\mathbb{C}^1=\mathrm{cl}(C\cup\bigcup\{\mathfrak{C}_x:x\in D\})$. Clearly $\mathbb{C}^1\subseteq\mathfrak{A}$ but $\mathfrak{A}\in\mathscr{A}(F)$ so there exists a finite $\mathbb{C}^2$ in F with $\mathbb{C}^1\subseteq\mathbb{C}^2\subseteq\mathfrak{A}$. For $x\in D$, $\mathfrak{C}_x\subseteq\mathbb{C}^2$ so $\mathfrak{C}_x\cap\mathbb{C}^2=\mathfrak{C}_x$. But then $(\mathfrak{C}_x,\Phi(\mathfrak{C}_x))\cap(\mathbb{C}^2,\Phi(\mathbb{C}^2))=(\mathfrak{C}_x,\Phi(\mathfrak{C}_x)\cap\Phi(\mathbb{C}^2))$ and as Φ is a function (and therefore single valued), we have $\Phi(\mathfrak{C}_x)\cap\Phi(\mathbb{C}^2)=\Phi(\mathfrak{C}_x)$, so $\Phi(\mathfrak{C}_x)\subseteq\Phi(\mathbb{C}^2)$. But $x\in\Phi(\mathfrak{C}_x)$ so $x\in\Phi(\mathbb{C}^2)$. We also have $\Phi(\mathbb{C}^2)\subseteq\mathfrak{B}$ since $\mathbb{C}^2$ is finite by the definition of $\mathfrak{B}$. So we now have a finite $\mathbb{C}^2$ such that

$$(C,D)\subseteq(\mathbb{C}^2,\Phi(\mathbb{C}^2))\subseteq(\mathfrak{A},\mathfrak{B})$$

and therefore $(\mathfrak{A},\mathfrak{B})\in\mathscr{A}(G)=\Phi$.

So indeed $\mathfrak{A}\in\mathbb{D}=\mathscr{A}(F)$.

(iii) Suppose $x\in\Phi(\mathfrak{A})$ for some $\mathfrak{A}\in\mathbb{D}$. Then $(\emptyset,\{x\})\subseteq(\mathfrak{A},\Phi(\mathfrak{A}))$ so $C_G(\emptyset,\{x\})$ is defined. Call it $(\mathfrak{X},\Phi(\mathfrak{X}))$. We claim $\mathfrak{X}$ is the required pseudo-inverse of x.

Suppose $\mathfrak{B}\in\mathbb{D}$ then we have the following chain of implications.
$\mathfrak{X}\subseteq\mathfrak{B}$ implies $\mathfrak{X}\cap\mathfrak{B}=\mathfrak{X}$
implies $(\mathfrak{X},\Phi(\mathfrak{X}))\cap(\mathfrak{B},\Phi(\mathfrak{B}))=(\mathfrak{X},\Phi(\mathfrak{X})\cap\Phi(\mathfrak{B}))$
implies $\Phi(\mathfrak{X})\cap\Phi(\mathfrak{B})=\Phi(\mathfrak{X})$ since Φ is single valued
implies $\Phi(\mathfrak{X})\subseteq\Phi(\mathfrak{B})$
implies $x\in\Phi(\mathfrak{B})$ since $x\in\Phi(\mathfrak{X})$ by construction.

Conversely, $x\in\Phi(\mathfrak{B})$ implies $(\emptyset,\{x\})\subseteq(\mathfrak{B},\Phi(\mathfrak{B}))$
implies $(\mathfrak{B},\Phi(\mathfrak{B}))\supseteq C_G(\emptyset,\{x\})=(\mathfrak{X},\Phi(\mathfrak{X}))$
implies $\mathfrak{B}\supseteq\mathfrak{X}$.

This completes the proof. $\square$

In the next section we strengthen this result to "a recursive frame map is a map in a *recursive* frame".

23. Recursive Frame Maps are Recursive Map Frames

Definition 23.1. A frame map Φ with domain $\mathbb{D}$ is said to be *recursive* if

(i) $\mathbb{D} = \mathscr{A}(F)$ for some recursive frame F, and

(ii) There is a partial recursive function $\ulcorner\Phi\urcorner$ whose domain is $\{\ulcorner\mathfrak{A}\urcorner : \mathfrak{A} \in F\}$ and is such that $\ulcorner\Phi\urcorner(\ulcorner\mathfrak{A}\urcorner) = \ulcorner\Phi(\mathfrak{A})\urcorner$ for all $\mathfrak{A} \in F$ where all indices are explicit.

Lemma 23.2. *Let $\Phi: \mathbb{D} \to \mathbb{C}$ be a recursive frame map with $\mathbb{D} = \mathscr{A}(F)$ where F is a recursive frame then*

(i) $\{\ulcorner\Phi(\mathfrak{A})\urcorner : \mathfrak{A} \in F\}$ *and* $\bigcup\{|\Phi(\mathfrak{A})| : \mathfrak{A} \in F\}$ *are recursively enumerable,*

(ii) *the function* $\ulcorner\Phi^{\leftarrow}\urcorner$ *with domain* $\bigcup\{|\Phi(\mathfrak{A})| : \mathfrak{A} \in F\}$ *defined by* $\ulcorner\Phi^{\leftarrow}\urcorner(x) = \ulcorner\Phi^{\leftarrow}(x)\urcorner$ *is partial recursive.*

Proof. (i) is immediate from Definition 23.1.

(ii) First, since F is a recursive frame for $\mathfrak{B} \in F$ we can enumerate, uniformly in $\mathfrak{B}$, an explicit list of all $\mathfrak{C} \in F$ with $\mathfrak{C} \subseteq \mathfrak{B}$.

Secondly,

$$\Phi^{\leftarrow}(x) = \bigcap\{\mathfrak{C} : x \in \Phi(\mathfrak{C}) \& \mathfrak{C} \in F\}$$
$$= \mathfrak{B} \cap \bigcap\{\mathfrak{C} : x \in \Phi(\mathfrak{C}) \& \mathfrak{C} \in F\}$$
$$= \bigcap\{\mathfrak{C} : \mathfrak{C} \subseteq \mathfrak{B} \& x \in \Phi(\mathfrak{C}) \& \mathfrak{C} \in F\}.$$

But this last is a finite intersection if $\mathfrak{B}$ is finite. So to compute $\Phi^{\leftarrow}(x)$ first enumerate F until a finite $\mathfrak{B}$ such that $x \in \Phi(\mathfrak{B})$ is found (and such always exists since Φ is continuous). Then compute the finite intersection above. $\square$

We note the following lemma since the technique is the same as for 23.2.

Lemma 23.3. *Let Φ be a recursive frame map. Then if $\mathfrak{A}$ is Dedekind and $\Phi(\mathfrak{A})$ is defined, $\Phi(\mathfrak{A})$ is Dedekind.*

Proof. Suppose $\Phi(\mathfrak{A})$ contains a recursively enumerable set B. Then $\Phi^{\leftarrow}(x) \subseteq \mathfrak{A}$ for all $x \in B$. Hence if dom $\Phi = \mathscr{A}(F)$ where F is a recursive frame then $\mathfrak{B} = C_F \cup \{\Phi^{\leftarrow}(x) : x \in B\} \subseteq \mathfrak{A}$. But $\mathfrak{A}$ is Dedekind and $\mathfrak{B}$ is recursively enumerable, so $\mathfrak{B}$, and therefore B, is finite and $\Phi(\mathfrak{A})$ is Dedekind. $\square$

Now we have the recursive version of Theorem 22.5.

Theorem 23.4. *Suppose $\Phi: \mathbb{D} \to \mathbb{C}^2$ where $\mathbb{D} \subseteq \mathbb{C}^1$. Then Φ is a recursive frame map if, and only if, there is a recursive frame $G \subseteq {}^{\circ}\mathbb{C}^1 \times {}^{\circ}\mathbb{C}^2$ such that $\Phi = \mathscr{A}(G)$.*

Proof. We augment the proof of Theorem 22.5 and use the same notation. Suppose Φ is a recursive frame map and dom $\Phi = \mathbb{D} = \mathscr{A}(F)$. Then $(C, D) \in G^*$ if, and only if,

(i) $D \subseteq \bigcup \{\Phi(\mathfrak{B}): \mathfrak{B} \in F\}$ and

(ii) $C \cup \bigcup \{\Phi^-(x): x \in D\} \in F^*$.

By Lemma 23.2 and the fact that F is a recursive frame it follows that $\ulcorner G^* \urcorner$ is recursively enumerable.

For condition (ii) of Definition 21.3 we use the fact that

$$C_G(C, D) = (\mathfrak{B}, \Phi(\mathfrak{B}))$$

where $\mathfrak{B} = C_F(C \cup \bigcup \{\Phi^-(x): x \in D\}$ from which we get that $\ulcorner \Phi \urcorner$ is partial recursive since $\ulcorner C_F \urcorner$, $\ulcorner \Phi^- \urcorner$ and the operation of taking the supremum of a finite number of systems are partial recursive.

Conversely, suppose G is a recursive frame and $\Phi = \mathscr{A}(G)$. Then $A \in F^*$ if, and only if, $(A, \emptyset) \in G^*$ so condition (i) of Definition 21.3 for G implies 21.3(i) for F. If $A \in F^*$ and $C_G(A, \emptyset) = (\mathfrak{A}^0, \mathfrak{A}^1)$, then $C_F(A) = \mathfrak{A}^0$ so conditions (i) and (ii) of Definition 21.3 for G give 21.3(ii) for F. So F is a recursive frame and condition (i) of Definition 23.1 is satisfied.

Condition (ii) of Definition 23.1 is also satisfied since we can compute $\ulcorner \Phi(\mathfrak{A}) \urcorner$ using $C_G(\mathfrak{A}, \emptyset) = (\mathfrak{A}, \Phi(\mathfrak{A}))$ and the fact that $\ulcorner C_G \urcorner$ is partial recursive by Definition 21.3(ii). (Note that we are using the fact here that $\ulcorner C_G \urcorner$ partial recursive implies we can compute explicit indices of all co-ordinates of $C_G(\mathfrak{A}, \emptyset)$ for finite $\mathfrak{A}$, in particular the last one, that is $\ulcorner \Phi(\mathfrak{A}) \urcorner$.) □

We close this section with a simple technical result we shall use later.

Theorem 23.5. *Suppose* F, G *are (recursive) frames with* $F \subseteq G$. *Let* $\Phi: \mathscr{A}(G) \to \mathbb{C}^2$ *be a (recursive) frame map and let* Ψ *be* Φ *restricted to* $\mathscr{A}(F)$. *Then* Ψ *is a (recursive) frame map.*

Proof. If $x \in \Psi(\mathfrak{A})$ then $x \in \Phi(\mathfrak{A})$ so $\Phi^-(x) \subseteq \mathfrak{A}$. Now $\mathfrak{A} \in \mathscr{A}(F)$ so $\Phi^-(x) \in F^*$ and $\Phi^-(x)$ is finite. Therefore $C_F \Phi^-(x) \subseteq \mathfrak{A}$ and $C_F \Phi^-(x)$ is also finite. Set $\Psi^-(x) = C_F \Phi^-(x)$. We already have $x \in \Psi(\mathfrak{A})$ implies $\Psi^-(x) \subseteq \mathfrak{A}$. For the converse suppose $\Psi^-(x) \subseteq \mathfrak{A}$ and $\mathfrak{A} \in \mathscr{A}(F)$, then $\Phi^-(x) \subseteq \mathfrak{A}$ so $x \in \Phi(\mathfrak{A})$. But $\Phi(\mathfrak{A}) = \Psi(\mathfrak{A})$ since $\mathfrak{A} \in \mathscr{A}(F)$ so $x \in \Psi(\mathfrak{A})$.

The recursive form of the result is immediate since the restriction of a partial recursive function to a recursively enumerable set is partial recursive. □

Theorem 23.6. *ſ* $\Phi: \mathbb{D}' \to \mathbb{D}$ *and* $\Psi: \mathbb{D}'' \to \mathbb{C}'$ *are recursive frame maps where* $\mathbb{C}' \subseteq \mathbb{D}'$, *then* $\Phi \circ \Psi$ *is a recursive frame map. (This generalizes to the case of several variables.)*

Proof. Exercise. □

24. Chains and Chain Types

An *n-chain* (in a category $\mathbb{C}$) is an *n* element sequence of morphisms of the form

$$\mathfrak{A}^0 \xrightarrow{p^1} \mathfrak{A}^1 \xrightarrow{p^2} \cdots \xrightarrow{p^n} \mathfrak{A}^n.$$

An ω-chain is just an *n*-chain for some *n*.

Definition 24.1. If *n* is finite, two *n*-chains

$$\mathfrak{A}^0 \xrightarrow{p^1} \mathfrak{A}^1 \longrightarrow \cdots \longrightarrow \mathfrak{A}^n$$

and

$$\mathfrak{B}^0 \xrightarrow{q^1} \mathfrak{B}^1 \longrightarrow \cdots \longrightarrow \mathfrak{B}^n$$

are said to be *n-equivalent* if there are isomorphisms $r^0, \ldots, r^n \in \mathbb{C}$ such that the following diagram commutes:

$$
\begin{array}{ccccccc}
\mathfrak{A}^0 & \xrightarrow{p^1} & \mathfrak{A}^1 & \longrightarrow \cdots \longrightarrow & \xrightarrow{p^n} & \mathfrak{A}^n \\
\downarrow{\scriptstyle r^0} & & \downarrow{\scriptstyle r^1} & & & \downarrow{\scriptstyle r^n} \\
\mathfrak{B}^0 & \xrightarrow{q^1} & \mathfrak{B}^1 & \longrightarrow \cdots \longrightarrow & \xrightarrow{q^n} & \mathfrak{B}^n.
\end{array}
$$

Clearly *n*-equivalence is an equivalence relation. We call the equivalence class of

$$\mathfrak{A}^0 \xrightarrow{p^1} \mathfrak{A}^1 \longrightarrow \cdots \longrightarrow \mathfrak{A}^n$$

the *n-chain type* of

$$\mathfrak{A}^0 \xrightarrow{p^1} \mathfrak{A}^1 \longrightarrow \cdots \longrightarrow \mathfrak{A}^n.$$

If $\mathfrak{A}^n$ is finite (or equivalently, if every $\mathfrak{A}^i$ for $0 \leq i \leq n$ is finite) then the *n*-chain type is said to be *finite*.

The 0-chain types are the isomorphism types of objects in $\mathbb{C}$. The 1-chain types are equivalence classes of morphisms in $\mathbb{C}$ and we call these *morphism types*. The *n*-chain types for all *n* constitute the *ω-chain types*.

Before we apply chain types to RETs we illustrate in the classical case how finite *ω-chain* types can be used to approximate infinite types. An $(\omega+1)$-*chain* is an infinite sequence of morphisms

$$\mathfrak{A}^0 \xrightarrow{p^1} \mathfrak{A}^1 \xrightarrow{p^2} \mathfrak{A}^2 \longrightarrow \cdots.$$

An $(\omega+1)$-chain is said to be *finite* if all its morphisms are finite.

Let

$$\mathfrak{A}^{\rightarrow} = \mathfrak{A}^0 \xrightarrow{p^1} \mathfrak{A}^1 \longrightarrow \cdots$$

and

$$\mathfrak{B}^{\rightarrow} = \mathfrak{B}^0 \xrightarrow{q^1} \mathfrak{B}^1 \longrightarrow \cdots$$

be two $(\omega+1)$-chains of finite morphisms of $\mathbb{C}$. Then we say that $\mathfrak{A}^{\rightarrow}$ and $\mathfrak{B}^{\rightarrow}$ (are $(\omega+1)$-*equivalent* and) have the same $(\omega+1)$-*chain type* if there are isomorphisms $\{r^i\colon i=0,1,2,\ldots\}$ in $\mathbb{C}$ such that the following diagram commutes:

$$
\begin{array}{ccccccccc}
\mathfrak{A}^0 & \xrightarrow{p^1} & \mathfrak{A}^1 & \longrightarrow & \cdots & \xrightarrow{p^n} & \mathfrak{A}^n & \xrightarrow{p^{n+1}} & \cdots \\
\downarrow{\scriptstyle r^0} & & \downarrow{\scriptstyle r^1} & & & & \downarrow{\scriptstyle r^n} & & \\
\mathfrak{B}^0 & \xrightarrow{q^1} & \mathfrak{B}^1 & \longrightarrow & \cdots & \xrightarrow{q^n} & \mathfrak{B}^n & \xrightarrow{q^{n+1}} & \cdots.
\end{array}
$$

Theorem 24.2. *Let*

$$\mathfrak{A}^{\rightarrow} = \mathfrak{A}^0 \xrightarrow{p^1} \mathfrak{A}^1 \longrightarrow \cdots$$

and

$$\mathfrak{B}^{\rightarrow} = \mathfrak{B}^0 \xrightarrow{q^1} \mathfrak{B}^1 \longrightarrow \cdots$$

be finite $(\omega+1)$-chains of morphisms of $\mathbb{C}$ such that for each n

$$\mathfrak{A}^0 \xrightarrow{p^1} \cdots \xrightarrow{p^n} \mathfrak{A}^n$$

and

$$\mathfrak{B}^0 \xrightarrow{q^1} \cdots \xrightarrow{q^n} \mathfrak{B}^n$$

have the same n-chain type. Then $\mathfrak{A}^{\rightarrow}$ and $\mathfrak{B}^{\rightarrow}$ have the same $(\omega+1)$-chain type.

Proof. (This is a special case of König's Lemma.) By assumption all the $\mathfrak{A}^i$ and $\mathfrak{B}^i$ are finite. Hence for each n there is only a finite set I^n of $(n+1)$-tuples of isomorphisms $(r^0,\ldots,r^n)$ such that

$$
\begin{array}{ccccccc}
\mathfrak{A}^0 & \xrightarrow{p^1} & \mathfrak{A}^1 & \longrightarrow & \cdots & \mathfrak{A}^n \\
\downarrow{\scriptstyle r^0} & & \downarrow{\scriptstyle r^1} & & & \downarrow{\scriptstyle r^n} \\
\mathfrak{B}^0 & \xrightarrow{q^1} & \mathfrak{B}^1 & \longrightarrow & \cdots & \mathfrak{B}^n
\end{array}
$$

commutes. Since I^0 is finite we can choose an r^0 in I^0 such that for infinitely many m we have $(r^0, s^1, \ldots, s^m) \in I^m$ for some $s^1, \ldots, s^m$. Suppose now $r^0, \ldots, r^n$ have been defined then again we choose r^{n+1} such that for infinitely many m we have $(r^0, r^1, \ldots, r^{n+1}, s^{n+2}, \ldots, s^{n+m}) \in I^{n+m}$ for some $s^{n+2}, \ldots, s^{n+m}$. But then

$$
\begin{array}{ccccccccc}
\mathfrak{A}^0 & \dashrightarrow{p^1} & \mathfrak{A}^1 & \longrightarrow & \cdots & \xrightarrow{p^n} & \mathfrak{A}^n & \longrightarrow & \cdots \\
\downarrow{\scriptstyle r^0} & & \downarrow{\scriptstyle r^1} & & \cdots & & \downarrow{\scriptstyle r^n} & & \cdots \\
\mathfrak{B}^0 & \xrightarrow{q^1} & \mathfrak{B}^1 & \longrightarrow & \cdots & \xrightarrow{q^n} & \mathfrak{B}^n & \longrightarrow & \cdots
\end{array}
$$

commutes. $\square$

Corollary 24.3. *Suppose*

$$\mathfrak{A}^0 \subseteq \mathfrak{A}^1 \subseteq \cdots \subseteq \mathfrak{A}^n \subseteq \cdots,$$

$$\mathfrak{B}^0 \subseteq \mathfrak{B}^1 \subseteq \cdots \subseteq \mathfrak{B}^n \subseteq \cdots,$$

$\mathfrak{A} = \bigcup \mathfrak{A}^i$, $\mathfrak{B} = \bigcup \mathfrak{B}^i$ *and all the* $\mathfrak{A}^i$, $\mathfrak{B}^i$ *are finite. Suppose that for each n there is an isomorphism* p^n: $\mathfrak{A}^n \to \mathfrak{B}^n$ *such that, writing* $p^{(m)}$ *for* $p^n | \mathfrak{A}^m$, $p^{(0)}$: $\mathfrak{A}^0 \cong \mathfrak{B}^0, \ldots, p^{(n-1)}$: $\mathfrak{A}^{n-1} \cong \mathfrak{B}^{n-1}$. *Then* $\mathfrak{A}$ *is isomorphic to* $\mathfrak{B}$. ☐

In the next section we shall extend this technique to approximate infinite RETs by using finite objects and also show how to extend relations on finite objects to relations on RETs. However, we first make the observation that although it may appear that *n*-chain types (especially when $n = \omega + 1$) take us right out of the category, for our purposes this does not happen. Each $(\omega + 1)$-chain type represents a dissection of an object in $\mathbb{C}$ into an increasing sequence of finite objects for, as the next results show, chain types do not distinguish between morphisms and inclusions.

Lemma 24.4. *Let* p: $\mathfrak{A} \to \mathfrak{A}'$ *be a finite morphism and* r: $\mathfrak{A} \to \mathfrak{B}$ *be a finite isomorphism, then there exists* $\mathfrak{B}' \supseteq \mathfrak{B}$ *and an isomorphism* r': $\mathfrak{A}' \to \mathfrak{B}'$ *such that the following diagram commutes.*

$$
\begin{array}{ccc}
\mathfrak{A} & \xrightarrow{\ p\ } & \mathfrak{A}' \\
\downarrow{\scriptstyle r} & & \downarrow{\scriptstyle r'} \\
\mathfrak{B} & \subseteq & \mathfrak{B}'.
\end{array}
$$

Proof. We can factor *p* through its image so we have the following commutative diagram

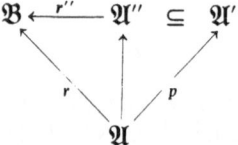

where $\mathfrak{A}'' = p(\mathfrak{A})$. But r'' is an isomorphism. By the Duplication Lemma 5.5 there exists $\mathfrak{B}' \supseteq \mathfrak{B}$ and an extension r' of r'' such that

$$
\begin{array}{ccc}
\mathfrak{A}'' & \subseteq & \mathfrak{A}' \\
\downarrow{\scriptstyle r} & & \downarrow{\scriptstyle r'} \\
\mathfrak{B} & \subseteq & \mathfrak{B}'
\end{array}
$$

commutes. Putting the two diagrams together gives the required result. ☐

Theorem 24.5. *Let $0 \leq n \leq \omega + 1$ and let*

$$\mathfrak{A}^{\rightarrow} = \mathfrak{A}^0 \xrightarrow{\ p^1\ } \mathfrak{A}^1 \xrightarrow{\ p^2\ } \cdots$$

be an n-chain of finite morphisms. Then there is an n-chain

$$\mathfrak{B}^{\rightarrow} = \mathfrak{B}^0 \subseteq \mathfrak{B}^1 \subseteq \cdots$$

of the same n-chain type as $\mathfrak{A}^{\rightarrow}$.

Proof. Let $\mathfrak{B}^0 = \mathfrak{A}^0$. Suppose isomorphisms $r^0, r^1, \ldots, r^m$ and objects $\mathfrak{B}^0, \mathfrak{B}^1, \ldots, \mathfrak{B}^m$ have been chosen so that

$$
\begin{array}{ccccccc}
\mathfrak{A}^0 & \xrightarrow{\ p^1\ } & \mathfrak{A}^1 & \xrightarrow{\ p^2\ } & \cdots & \mathfrak{A}^m \\
\downarrow{\scriptstyle r^0} & & \downarrow{\scriptstyle r^1} & & & \downarrow{\scriptstyle r^m} \\
\mathfrak{B}^0 & \subseteq & \mathfrak{B}^1 & \subseteq & \cdots & \mathfrak{B}^m
\end{array}
$$

commutes. By the lemma, since p^m, $\mathfrak{A}^r$, $\mathfrak{B}^r$ are finite, there exists $\mathfrak{B}^{m+1} \supseteq \mathfrak{B}^m$ and an isomorphism $r^{m+1} \colon \mathfrak{A}^{m+1} \to \mathfrak{B}^{m+1}$ such that

$$
\begin{array}{ccc}
\mathfrak{A}^m & \xrightarrow{\ p^{m+1}\ } & \mathfrak{A}^{m+1} \\
\downarrow{\scriptstyle r^m} & & \downarrow{\scriptstyle r^{m+1}} \\
\mathfrak{B}^m & \subseteq & \mathfrak{B}^{m+1}
\end{array}
$$

commutes. Put the two diagrams together. The result now follows by induction. $\square$

Note that the object $\mathfrak{B} = \bigcup \{\mathfrak{B}^i \colon i < n\}$ is the dissected object referred to above. $\mathfrak{B}$ is determined uniquely up to isomorphism by the $(\omega + 1)$-chain type.

25. Extending Relations Using Frames

In this section we first show how frames arise naturally from solutions of equations involving combinatorial functors. Then we reverse the process to extend relations from finite things (in this case finite chain types) to RETs. It will then be possible to characterize for some categories those functions which, when considered as relations, extend to functions on RETs.

Theorem 25.1. *Let $G, H \colon \mathbb{C}^1 \to \mathbb{C}^2$ be finitary recursive combinatorial functors and let p be a one-one partial recursive function. Let*

$$F = \{\mathfrak{A} \in {}^{\circ}\mathbb{C}^1 \colon p|G(\mathfrak{A}) \text{ is a } \mathbb{C}^2\text{-morphism of } G(\mathfrak{A}) \text{ onto } H(\mathfrak{A})\}.$$

Then F *is a recursive frame. Moreover, if* $\mathfrak{A}$ *is Dedekind then* $p|G(\mathfrak{A})$ *is a* $\mathbb{C}^2$*-morphism of* $G(\mathfrak{A})$ *onto* $H(\mathfrak{A})$ *if, and only if,* $\mathfrak{A}$ *is attainable from* F.

Proof. First we claim that F is a frame.

Suppose $\{\mathfrak{B}^i: i \in I\} \subseteq F$. Now

$$p\, G(\mathfrak{B}^i) = H(\mathfrak{B}^i)$$

so

$$\bigcap p\, G(\mathfrak{B}^i) = \bigcap H(\mathfrak{B}^i)$$

$$= H(\bigcap \mathfrak{B}^i) \qquad \text{since } H \text{ preserves intersections}.$$

But

$$\bigcap p\, G(\mathfrak{B}^i) = p \cap G(\mathfrak{B}^i) \qquad \text{since } p \text{ is one-one}$$

$$= p\, G(\bigcap \mathfrak{B}^i) \qquad \text{since } G \text{ preserves intersections}.$$

Therefore $p\, G(\bigcap \mathfrak{B}^i) = H(\bigcap \mathfrak{B}^i)$ and F is a frame.

Next we show that F is a recursive frame. We use a short cut available to us because p is partial recursive to show F* is recursively enumerable. In fact it suffices to show F is recursively enumerable, for $A \in$ F* if and only if there exists $\mathfrak{A} \in$ F with $A \subseteq |\mathfrak{A}|$.

To enumerate F simply enumerate those finite $\mathfrak{A}$ such that $p|G(\mathfrak{A})$ is a $\mathbb{C}^2$-morphism of $G(\mathfrak{A})$ onto $H(\mathfrak{A})$. Since G, H are finitary recursive this process is indeed effective.

To compute $C_F(A)$ we give an inductive definition which effectively yields $C_F(A)$ for $A \in$ F*.

Let $\mathfrak{B}$ be the smallest finite set (which exists for $A \in$ F*) satisfying 1, 2, 3 below. Then $\mathfrak{B} = C_F(A)$.

1. $A \subseteq \mathfrak{B}$,
2. $\mathrm{cl}(\mathfrak{B}) = \mathfrak{B}$,
3. $p|G(\mathfrak{B})$ is a $\mathbb{C}^2$-isomorphism of $G(\mathfrak{B})$ onto $H(\mathfrak{B})$.

Hence F is a recursive frame.

Finally, suppose $p\colon G(\mathfrak{A}) \simeq H(\mathfrak{A})$ and $\mathfrak{A}$ is Dedekind. Let $A \subseteq \mathfrak{A}$ and let $\mathfrak{B}$ be the smallest system satisfying 1, 2, 3 above. Then since $\mathfrak{A}$ is Dedekind, $A \subseteq \mathfrak{A}$ and generating $\mathfrak{B}$ from A is effective, $\mathfrak{B}$ is a recursively enumerable subset of $\mathfrak{A}$, so $\mathfrak{B}$ is finite. Hence $\mathfrak{B} = C_F(\mathfrak{A}) \in$ F. Since A was arbitrary, $\mathfrak{A}$ is attainable from F.

Conversely, suppose $\mathfrak{A} \in \mathscr{A}(F)$.

Let $A \subseteq \mathfrak{A}$ then

$$p\colon G(C_F(A)) \simeq H(C_F(A)).$$

But $\mathfrak{A} = \bigcup \{C_F(A): A \text{ is finite } \& A \subseteq \mathfrak{A}\}$ is a directed union and all the relations in $\mathfrak{A}$ have only a finite number of arguments hence

$$p\colon G(\mathfrak{A}) \simeq H(\mathfrak{A}).$$

This completes the proof. $\square$

If $p^i = p|G(\mathfrak{A}^i)$,

$$\mathfrak{A}^0 \subseteq \mathfrak{A}^1 \subseteq \cdots \subseteq \mathfrak{A}^n$$

is an n-chain of inclusions and each $\mathfrak{A}^i \in F$ then

$$
\begin{array}{ccccc}
G(\mathfrak{A}^0) & \longrightarrow & G(\mathfrak{A}^1) & \longrightarrow \cdots & G(\mathfrak{A}^n) \\
\downarrow p^0 & & \downarrow p^1 & & \downarrow p^n \\
H(\mathfrak{A}^0) & \longrightarrow & H(\mathfrak{A}^1) & \longrightarrow \cdots & H(\mathfrak{A}^n)
\end{array}
$$

commutes. Though a trivial observation here, this is central for the next development.

Definition 25.2. Let R be a set of finite ω-chain types of $\mathbb{C}$ and F be a frame in $\mathbb{C}$. Then F is said to be an ^nR-*frame* where $0 \leq n \leq \omega$ if, whenever $\mathfrak{A}^0, \mathfrak{A}^1, \ldots, \in F$ and

$$\mathfrak{A}^{\rightarrow} = \mathfrak{A}^0 \subseteq \mathfrak{A}^1 \subseteq \cdots$$

is the n-chain of inclusion maps then the chain type of $\mathfrak{A}^{\rightarrow}$ is in R.

Note that F is a ^0R-frame if $\mathfrak{A} \in F$ implies the type of $\mathfrak{A}$, that is $\langle \mathfrak{A} \rangle$, is in R.

Definition 25.3. Let R be a set of finite ω-chain types of $\mathbb{C}$. Then the n-*chain extension* of R to the Dedekind $\mathbb{C}$-RETs, written $\mathbb{C}(^nR)$, is the set of all $X \in \Lambda(\mathbb{C})$ such that for some $\mathfrak{X} \in X$, $\mathfrak{X}$ is attainable from some recursive ^nR-frame. We write $\mathbb{C}(R)$ for $\mathbb{C}(^0R)$.

Lemma 25.4. *Let* F *be a recursive frame in* $\mathbb{C}$ *and let* p *be a one-one partial recursive function. Let*

$$G = \{p(\mathfrak{A}): \mathfrak{A} \in F \ \& \ p|\mathfrak{A} \ \text{is a} \ \mathbb{C}\text{-morphism of} \ \mathfrak{A} \ \text{onto} \ p(\mathfrak{A})\}.$$

Then G *is a recursive frame. If* $\mathfrak{A} \in \mathscr{A}(F)$ *and* $p|\mathfrak{A}$ *is a* $\mathbb{C}$-*morphism of* $\mathfrak{A}$ *onto* $p(\mathfrak{A})$ *then* $p(\mathfrak{A}) \in \mathscr{A}(G)$.

Proof. Since p is one-one $p(\cap \mathfrak{A}^i) = \cap p(\mathfrak{A}^i)$ so G is a frame. Clearly G is recursive since $A \in F^*$ implies A is finite, so we can verify if $p|C_F(A)$ is a $\mathbb{C}$-morphism and therefore $G^* = \{p(A): A \in F^* \& p|C_F(A) \text{ is a } \mathbb{C}\text{-morphism}\}$ is recursively enumerable. $C_G(A) = p \, C_F \, p^{-1}(A)$ means $\ulcorner C_G \urcorner$ is partial recursive since, when given an explicit index for a finite set A contained in the range of p, we can compute an explicit index of $p^{-1}(A)$; and similarly for $p(B)$ when $B \subseteq \text{dom } p$. $\square$

Theorem 25.5. *If* $Y \in \mathbb{C}(^nR)$ *and* $Y = \langle \mathfrak{Y} \rangle$ *then* $\mathfrak{Y}$ *is attainable from some recursive* ^nR-*frame.*

Proof. If $Y \in \mathbb{C}(^n R)$ then there is an $\mathfrak{X} \in Y$ and a recursive $^n R$-frame F such that $\mathfrak{X} \in \mathscr{A}(F)$. Suppose $p: \mathfrak{X} \simeq \mathfrak{Y}$, then by Lemma 25.4, $p(F)$ is an $^n R$-frame from which $\mathfrak{Y}$ is attainable. $\square$

So the extensions are independent of the representatives of the RETs.

We observe that a finitary combinatorial functor $F: \mathbb{C}^1 \to \mathbb{C}^2$ induces not only a map on finite isomorphism types but also a map F on ω-chain types induced by mapping

$$\mathfrak{A}^0 \xrightarrow{p^1} \mathfrak{A}^1 \xrightarrow{p^2} \cdots \mathfrak{A}^n$$

to

$$F(\mathfrak{A}^0) \xrightarrow{F(p^1)} F(\mathfrak{A}^1) \xrightarrow{F(p^1)} \cdots F(\mathfrak{A}^n).$$

Since if

$$
\begin{array}{ccccc}
\mathfrak{A}^0 & \longrightarrow & \mathfrak{A}^1 & \longrightarrow \cdots & \mathfrak{A}^n \\
\downarrow{\scriptstyle r^0} & & \downarrow{\scriptstyle r^1} & & \downarrow{\scriptstyle r^n} \\
\mathfrak{B}^0 & \longrightarrow & \mathfrak{B}^1 & \longrightarrow \cdots & \mathfrak{B}^n
\end{array}
$$

commutes then

$$
\begin{array}{ccccc}
F(\mathfrak{A}^0) & \longrightarrow & F(\mathfrak{A}^1) & \longrightarrow \cdots & F(\mathfrak{A}^n) \\
\downarrow{\scriptstyle F(r^0)} & & \downarrow{\scriptstyle F(r^1)} & & \downarrow{\scriptstyle F(r^n)} \\
F(\mathfrak{B}^0) & \longrightarrow & F(\mathfrak{B}^1) & \longrightarrow \cdots & F(\mathfrak{B}^n)
\end{array}
$$

commutes and $F(r^i)$ is a $\mathbb{C}^2$-isomorphism if r^i is a $\mathbb{C}^1$-isomorphism.

Now if $\mathfrak{A} \in \mathbb{C}^1$ and $\langle \mathfrak{A} \rangle$ is its 0-chain type, then $F \langle \mathfrak{A} \rangle = \langle F \mathfrak{A} \rangle$. For 1-chain types we write $\langle p \rangle$ where p is a $\mathbb{C}^1$-morphism so $F \langle p \rangle = \langle F p \rangle$ for $\langle p \rangle$ a morphism type.

Now the finite isomorphism types are the same as the finite $\mathbb{C}$-RETs and if F is a finitary recursive combinatorial functor then $F \langle \mathfrak{A} \rangle$ as an induced function on finite 0-chain types and $F \langle \mathfrak{A} \rangle$ as an induced function on finite $\mathbb{C}$-RETs may be identified.

Corollary 25.6. *Let* $G, H: \mathbb{C}^1 \to \mathbb{C}^2$ *be finitary recursive combinatorial functors and let* $0 \leq n \leq \omega$. *Let* $^n R$ *be the set of finite n-chain type solutions* C *of* $G(C) = H(C)$. *If* $X \in \Lambda(\mathbb{C}^1)$ *and* $G(X) = H(X)$, *then* $X \in \mathbb{C}^1(^n R)$.

Proof. We use the proof of Theorem 25.1 and add a little more.

If $G(X) = H(X)$ and $\mathfrak{A} \in X$ then there is a one-one partial recursive function p such that $p | G(\mathfrak{A})$ is a $\mathbb{C}^2$-isomorphism of $G(\mathfrak{A})$ onto $H(\mathfrak{A})$. But then $\mathfrak{A} \in \mathscr{A}(F)$ (using the same notation as for Theorem 25.1) and F is an $^n R$-frame for any n, since, if $\mathfrak{A}^0 \subseteq \mathfrak{A}^1 \subseteq \cdots \subseteq \mathfrak{A}^n$ then $\mathfrak{A}^1, \ldots, \mathfrak{A}^n \in F$

gives that

$$G(\mathfrak{A}^0) \longrightarrow G(\mathfrak{A}^1) \longrightarrow \cdots G(\mathfrak{A}^n)$$

$$\Big\downarrow p \qquad\qquad \Big\downarrow p \qquad\qquad \Big\downarrow p$$

$$H(\mathfrak{A}^0) \longrightarrow H(\mathfrak{A}^1) \longrightarrow \cdots H(\mathfrak{A}^n)$$

commutes, so the n-chain type of $\mathfrak{A}^0 \subseteq \cdots \subseteq \mathfrak{A}^n$ is in ^nR.
    This completes the proof. $\square$

Corollary 25.7. *Let* $t^1 = t^2$ *be an equation built up from variables and function symbols for functions on RETs induced by finitary recursive combinatorial functors from* $\mathbb{C}^1$ *to* $\mathbb{C}^2$. *Let* $0 \leq n \leq \omega$ *and let R be the set of finite n-chain type solutions to* $t^1 = t^2$. *If* $X \in \varLambda(\mathbb{C}^1)$ *satisfies* $t^1 = t^2$ *then* $X \in \mathbb{C}^1(^nR)$.

Proof. Immediate from Corollary 25.6 and the closure of finitary recursive combinatorial functors under composition (*cf.* Theorem 15.7). $\square$

Part VII. The Dimension Case

26. Extensions of Solutions of Equations

We investigate converses to Corollary 25.6. We delineate relations R such that $\mathbb{C}(R)$ is the set of solutions to $F^1(X) = F^2(X)$ where F^1, F^2 are induced by finitary recursive combinatorial functors F^1, F^2. The neatest situation is that with strict combinatorial functors, dimension, and models with degree 1.

Theorem 26.1. *Let $\mathbb{C}$ be a suitable category, let $\mathbb{D} = \mathbb{C}(\mathfrak{M})$ be a suitable category with dimension, where $\mathfrak{M}$ has degree 1. Let F^1, $F^2 \colon \mathbb{C} \to \mathbb{D}$ be finitary recursive strict combinatorial functors induced by recursive precombinatorial operators G^1, G^2. Let R be the set of finite isomorphism type (that is, 0-chain type) solutions to $F^1(X) = F^2(X)$. Then*

$$\mathbb{C}(R) = \{ X \in \Lambda(\mathbb{C}) \colon F^1(X) = F^2(X) \} \, .$$

First we need some lemmata.

Lemma 26.2. *Let $F \colon \mathbb{C} \to \mathbb{D}$ be a finitary recursive strict combinatorial functor. Suppose $\mathbb{D}$ has dimension and F is induced by a recursive precombinatorial operator G. Then there is a general recursive function $\ulcorner G \urcorner$ assigning to each $\ulcorner \mathfrak{A} \urcorner$ with $\mathfrak{A} \in {}^\circ\mathbb{C}$ an explicit index $\ulcorner G(\mathfrak{A}) \urcorner$ of the finite set $G(\mathfrak{A})$.*

Proof. For any $\mathfrak{A} \in {}^\circ\mathbb{C}$ we can compute an explicit index of $F(\mathfrak{A})$ since F is finitary recursive. Now we can enumerate all $G(\mathfrak{A}')$ with $\mathfrak{A}' \subseteq \mathfrak{A}$ and $\mathfrak{A}' \in {}^\circ\mathbb{C}$, since G is recursive. But $F(\mathfrak{A})$ is finite and

$$F(\mathfrak{A}) = \mathrm{cl} \bigcup \{ G(\mathfrak{A}') \colon \mathfrak{A}' \subseteq \mathfrak{A} \ \& \ \mathfrak{A}' \in {}^\circ\mathbb{C} \}$$

where the union is disjoint. We compute this closure and at some finite stage identify it as all of $F(\mathfrak{A})$ from the explicit index of the latter. At this stage all of $G(\mathfrak{A})$ has been enumerated and its explicit index may be computed. $\quad\square$

Lemma 26.3. *Let $K \subseteq {}^\circ\mathbb{C}$ be a frame and let $F \colon \mathbb{C} \to \mathbb{D}$ be a strict combinatorial functor induced by a precombinatorial operator G.*

Set

$$H(\mathfrak{A}) = \bigcup \{G(\mathfrak{A}'): \mathfrak{A}' \in {}^{\circ}\mathbb{C} \,\&\, C_{\mathsf{K}}(\mathfrak{A}') = \mathfrak{A}\}$$

for $\mathfrak{A} \in \mathsf{K}$. *Then*

$$\bigcup \{G(\mathfrak{A}'): \mathfrak{A}' \in {}^{\circ}\mathbb{C} \,\&\, C_{\mathsf{K}}(\mathfrak{A}') \; exists\}$$

is the disjoint union of all $H(\mathfrak{A})$ *with* $\mathfrak{A} \in \mathsf{K}$. *Moreover, for all* $\mathfrak{A}$ *attainable from* K *we have*

$$F(\mathfrak{A}) = \mathrm{cl} \bigcup \{H(\mathfrak{A}'): \mathfrak{A}' \in \mathsf{K} \,\&\, \mathfrak{A}' \subseteq \mathfrak{A}\}.$$

Proof. Immediate from the definition of F from G and the definition of H from G and K. $\square$

Proof of Theorem 26.1. By Corollary 25.6 it suffices to show that if $\mathsf{X} \in \mathbb{C}(R)$, then $\mathsf{F}^1(\mathsf{X}) = \mathsf{F}^2(\mathsf{X})$. Suppose $\mathsf{X} \in \mathbb{C}(R)$, then there exists a Dedekind $\mathfrak{A}$ in X and a recursive frame $\mathsf{K} \subseteq {}^{\circ}\mathbb{C}$ such that $\mathfrak{A}$ is attainable from K and, for every $\mathfrak{B}$ in K, the 0-chain type of $\mathfrak{B}$ is in R. Using the precombinatorial operators G^1, G^2 which induce F^1, F^2 respectively define H^1, H^2 corresponding to F^1, F^2 as in Lemma 26.3. Then, for all $\mathfrak{B} \in \mathsf{K}$,

$$(*) \qquad \begin{aligned} &\sum \{\mathrm{card}\, H^1(\mathfrak{B}'): \mathfrak{B}' \subseteq \mathfrak{B} \,\&\, \mathfrak{B}' \in \mathsf{K}\} \\ &= \sum \{\mathrm{card}\, H^2(\mathfrak{B}'): \mathfrak{B}' \subseteq \mathfrak{B} \,\&\, \mathfrak{B}' \in \mathsf{K}\}, \end{aligned}$$

since the left-hand side is $\dim F^1(\mathfrak{B})$ and the right $\dim F^2(\mathfrak{B})$ by Lemma 26.3.

We now prove by induction on the inclusion ordering of K that $\mathrm{card}\, H^1(\mathfrak{B}) = \mathrm{card}\, H^2(\mathfrak{B})$ for all $\mathfrak{B} \in \mathsf{K}$. If $\mathfrak{A}^0$ is the least element of K then $(*)$ reduces to $\mathrm{card}\, H^1(\mathfrak{A}^0) = \mathrm{card}\, H^2(\mathfrak{A}^0)$. Now suppose $\mathfrak{B} \in \mathsf{K}$ and $\mathrm{card}\, H^1(\mathfrak{B}') = \mathrm{card}\, H^2(\mathfrak{B}')$ for all $\mathfrak{B}' \subset \mathfrak{B}$ with $\mathfrak{B}' \in \mathsf{K}$. Then in $(*)$ all terms on the left except $\mathrm{card}\, H^1(\mathfrak{B})$ are equal to the corresponding terms on the right. Hence $\mathrm{card}\, H^1(\mathfrak{B}) = \mathrm{card}\, H^2(\mathfrak{B})$.

Since $\mathbb{C}$ is suitable and K is a recursive frame we can effectively enumerate $H^1(\mathfrak{A})$, $H^2(\mathfrak{A})$ without repetitions, uniformly in $\mathfrak{A}$, for $\mathfrak{A} \in \mathsf{K}$. Let τ be the bijection which has domain $\bigcup \{H^1(\mathfrak{A}'): \mathfrak{A}' \in \mathsf{K}\}$ and codomain $\bigcup \{H^2(\mathfrak{A}'): \mathfrak{A}' \in \mathsf{K}\}$ and which maps each $H^1(\mathfrak{A}')$ onto $H^2(\mathfrak{A}')$ in order of enumeration. Since K is a recursive frame and G^1, G^2 are recursive precombinatorial operators, τ is partial recursive. But then $\mathrm{cl}\, \tau: \mathrm{cl\, dom}\, \tau \to \mathrm{cl\, codom}\, \tau$ is also partial recursive (by effectivizing Lemmata 5.2 and 6.9). Now by Lemma 26.3 the restriction $\tau_{\mathfrak{A}}: F^1(\mathfrak{A}) \to F^2(\mathfrak{A})$ of $\mathrm{cl}\, \tau$ is an isomorphism of $\mathbb{D}$ since $\mathfrak{A}$ is attainable from K, so $\langle F^1(\mathfrak{A}) \rangle = \langle F^2(\mathfrak{A}) \rangle$ and this completes the proof. $\square$

We remark that (as in Theorem 7.6) we could formulate this result as an (effective) natural equivalence between the restrictions of F^1 and F^2 to inclusion maps between objects in $\mathscr{A}(\mathsf{K})$.

27. Universal Horn Sentences

We extend the techniques and results of Section 26 to implications of the form

$$t^1 = t^2 \& \cdots \& t^{2k-1} = t^{2k} \rightarrow r = s.$$

This kind of formula when universally closed is a universal Horn sentence.

We need a lemma.

Lemma 27.1. *Suppose* F, G *are recursive frames in* $\mathbb{C}$. *Then* $F \cap G$ *is a recursive frame in* $\mathbb{C}$. *If* $\mathfrak{A}$ *is Dedekind then* $\mathfrak{A}$ *is attainable from* $F \cap G$ *if, and only if,* $\mathfrak{A}$ *is attainable from both* F *and* G.

Proof. Clearly $F \cap G$ is a frame.

We now show how to compute $C_{F \cap G}(A)$ for $A \in (F \cap G)^*$. Since this procedure terminates this will show that condition (ii) of 21.3 is also satisfied, so $F \cap G$ is a recursive frame.

For any A let $A^0 = A$, $A^1 = C_F(A^0)$, $A^2 = C_G(A^1)$, $A^3 = C_F(A^2)$ and so on. We claim $A \in (F \cap G)^*$ if, and only if, A^n is defined for all n and for sufficiently large n, $A^m = A^r$ for all $m, r \geq n$. Suppose $A \in (F \cap G)^*$ then $A \in F^*$ so A^1 is defined and

$$A \subseteq A^1 = C_F(A) \subseteq C_{F \cap G}(A).$$

By symmetry A^2 is defined and by induction A^n is defined for all n. Now $C_{F \cap G}(A)$ is defined so the A^n constitute a weakly increasing chain of subsets of the finite set $C_{F \cap G}(A)$. So A^n is eventually constant.

Conversely, suppose A^n is defined for all n and eventually constant and equal to $B = A^m = A^{m+1}$. Then $B = C_F(B) = C_G(B)$ and therefore $B = C_{F \cap G}(B)$. So $A \in (F \cap G)^*$.

Now $\ulcorner C_F \urcorner$ and $\ulcorner C_G \urcorner$ are partial recursive and $\ulcorner F \cap G \urcorner$ is recursively enumerable, so we can effectively generate the $A \in (F \cap G)^*$ and the corresponding A^n. But we can also determine when we have an m such that $A^m = A^{m+1}$. Hence $\ulcorner C_{F \cap G} \urcorner$ is partial recursive with domain $\ulcorner (F \in G)^* \urcorner$.

Now trivially $(F \cap G)^* \subseteq F^* \cap G^*$. But we have shown if $A \in (F \cap G)^*$ then $C_{F \cap G}(A)$ exists and is in $F \cap G$ so $A \in F^* \cap G^*$. Hence $(F \cap G)^*$ is recursively enumerable. Thus $F \cap G$ is indeed a recursive frame.

Clearly $\mathfrak{A} \in \mathscr{A}(F \cap G)$ implies $\mathfrak{A} \in \mathscr{A}(F) \cap \mathscr{A}(G)$. Suppose $\mathfrak{A}$ is Dedekind, A^0 is a finite subset of $\mathfrak{A}$ and $\mathfrak{A} \in \mathscr{A}(F) \cap \mathscr{A}(G)$. Then $A^0, A^1, \ldots$ are all defined and form an effectively enumerable weakly increasing chain of subsets of $|\mathfrak{A}|$. But $\mathfrak{A}$ is Dedekind so the chain is eventually constant, say $A^m = A^{m+1}$. But then

$$A^m = C_F(A^m) = C_G(A^m) \quad \text{so} \quad A^m = C_{F \cap G}(A^m)$$

and $A^0 \in (F \cap G)^*$. Therefore $\mathfrak{A} \in \mathscr{A}(F \cap G)$. This completes the proof. $\square$

Corollary 27.2. *Let R, S be relations on ω-chain types then for any n∈ω,*

(i) $\mathbb{C}(^n(R \cap S)) = \mathbb{C}(^nR) \cap \mathbb{C}(^nS)$ *and*

(ii) $^nR \subseteq {}^nS$ *implies* $\mathbb{C}(^nR) \subseteq \mathbb{C}(^nS)$.

Proof. Trivially (i) implies (ii). Since an $^n(R \cap S)$-frame is both an ^nR-frame and an ^nS-frame $\mathbb{C}(^n(R \cap S)) \subseteq \mathbb{C}(^nR) \cap \mathbb{C}(^nS)$.

Suppose then that $\mathfrak{A} \in \mathscr{A}(F)$ and $\mathfrak{A} \in \mathscr{A}(G)$ where F is a recursive ^nR-frame and G a recursive ^nS-frame. By Lemma 27.1, $F \cap G$ is a recursive frame and indeed an $^n(R \cap S)$-frame. But if $\mathfrak{A}$ is Dedekind then $\mathfrak{A} \in \mathscr{A}(F \cap G)$ so $\langle \mathfrak{A} \rangle \in \mathbb{C}(^n(R \cap S))$. That is, $\mathbb{C}(^nR) \cap \mathbb{C}(^nS) \subseteq \mathbb{C}(^n(R \cap S))$ and the corollary is established. ☐

Let $\mathbb{D} = \mathbb{C}(\mathfrak{M})$ be a suitable category with dimension, where $\mathfrak{M}$ has degree 1. Let $t^1, \ldots, t^{2k}$ be terms such that for each i, t^{2i-1}, t^{2i} both denote functions induced by (composition of) finitary recursive combinatorial functors from $\mathbb{C}$ into the same category $\mathbb{E}^i$ and r, s denote functions induced by finitary recursive strict combinatorial functors from $\mathbb{C}$ into $\mathbb{D}$. Note that we do not assume $\mathbb{C}$ or the $\mathbb{E}^i$ have dimension, only that $\mathbb{D}$ has dimension.

Theorem 27.3. *With the above hypothesis, if*

(1) $$t^1(\mathsf{X}) = t^2(\mathsf{X}) \,\&\, \ldots \,\&\, t^{2k-1}(\mathsf{X}) = t^{2k}(\mathsf{X}) \to r(\mathsf{X}) = s(\mathsf{X})$$

is true for finite isomorphism types X *then it is true for all effective Dedekind types* X *in* $\Lambda(\mathbb{C})$.

Proof. Let R^i be the set of all finite isomorphism types solving $t^{2i-1}(\mathsf{X}) = t^{2i}(\mathsf{X})$ and similarly let S be the set of all finite isomorphism type solutions of $r(\mathsf{X}) = s(\mathsf{X})$. By hypothesis $R^1 \cap \cdots \cap R^k \subseteq S$. By Corollary 27.2 since $\mathbb{C}(^0R^i)$ is $\mathbb{C}(R^i)$,

$$\mathbb{C}(R^1 \cap \cdots \cap R^k) = \mathbb{C}(R^1) \cap \cdots \cap \mathbb{C}(R^k)$$

and $R^1 \cap \cdots \cap R^k \subseteq S$ implies that since $\mathbb{C}(^0S)$ is $\mathbb{C}(S)$,

$$\mathbb{C}(R^1 \cap \cdots \cap R^k) \subseteq \mathbb{C}(S)$$

so $$\mathbb{C}(R^1) \cap \cdots \cap \mathbb{C}(R^k) \subseteq \mathbb{C}(S).$$

Now by Corollary 25.6, if $\mathsf{X} \in \Lambda(\mathbb{C})$ satisfies $t^1(\mathsf{X}) = t^2(\mathsf{X}) \,\&\, \ldots \,\&\, t^{2k-1}(\mathsf{X}) = t^{2k}(\mathsf{X})$ then $\mathsf{X} \in \mathbb{C}(R^1) \cap \cdots \cap \mathbb{C}(R^k)$. Therefore $\mathsf{X} \in \mathbb{C}(S)$. But then by Theorem 26.1, $r(\mathsf{X}) = s(\mathsf{X})$. This completes the proof. ☐

Corollary 27.4. *With the same hypotheses as Theorem 27.3. Suppose there is a natural number z such that whenever* $\mathsf{X} = (\mathsf{X}_0, \ldots, \mathsf{X}_{n-1})$ *is an* *n-tuple of finite isomorphism types with each* X_i *of cardinality* $\geq z$,

$$t^1(X) = t^2(X) \& \ldots \& t^{2k-1}(X) = t^{2k}(X) \rightarrow r(X) = s(X)$$

is true then, for all n-tuples of infinite Dedekind types the same implication holds.

Proof. (We use the notation in the proof of the theorem.) Let $\mathfrak{X} \in X$ then there is a recursive $^0(R^1 \cap \cdots \cap R^k)$-frame F such that $\mathfrak{X}$ is attainable from F. Since each $\mathfrak{X}_i$ is infinite there exists $\mathfrak{Y} \in F$, with card $\mathfrak{Y}_i \geq z$ for each i, such that $\mathfrak{Y} \subseteq \mathfrak{X}$. Let $F' = \{\mathfrak{Z} \in F: \mathfrak{Z} \supseteq \mathfrak{Y}\}$, then F' is a ^0S-frame since every $^0(R^1 \cap \cdots \cap R^k)$-frame is a ^0S-frame. F' is a recursive frame since F' is clearly recursively enumerable and $C_{F'}(\mathfrak{Z}) = C_F(\mathfrak{Z})$-restricted to F'. Clearly $\mathfrak{X} \in \mathscr{A}(F')$ so the result now follows from Theorem 26.1. $\square$

28. Universal Sentences I

This section and the section after this present the culmination of our results in the dimension case. In this section we determine sufficient, and in the next section (useful and practical) necessary, conditions for the truth in Dedekind $\mathbb{C}$-types of universal sentences involving finitary recursive strict combinatorial functors and equality when $\mathbb{C}$ has dimension. The method we use is, we believe, as simple as possible though not as general as might be. (The diligent reader may combine the results on extending relations obtained in succeeding sections and the argument paralleling Theorem 18.5 with these results to obtain an exact generalization of Nerode's [1961] Theorem 11.1.) The technique is an extension of Myhill's first note on combinatorial functions [1958] and of the earliest characterization by Myhill and Nerode of universal sentences true of $\mathbb{S}$-Dedekind types.

We use the technique of Section I to deal with the case of many variables. In this and the next section all categories have dimension and degree one.

We shall be thinking of equations interpreted in five ways, so, to avoid complication of notation, we shall not distinguish between the variables corresponding to the various categories, we shall merely assume that we have a many-sorted language with the variables being of the appropriate sort. We shall also suppress "k-tuples of" and simply write, for example, "natural numbers" instead of "k-tuples of natural numbers."

If $G: \mathbb{C} \rightarrow \mathbb{D}$ is a finitary recursive combinatorial functor and $\mathbb{C}, \mathbb{D}$ have dimension, then G induces a combinatorial function dim G given by

$$\dim G(\dim \mathfrak{A}) = \dim G(\mathfrak{A}).$$

Similarly if R is a relation on finite isomorphism types we set

$$\dim R = \{\dim \mathfrak{A} : \mathfrak{A} \in R\}.$$

We shall write $\Lambda^\infty(\mathbb{C})$ for the set of all infinite Dedekind $\mathbb{C}$-types. Analogously we write $\Lambda^{\mathrm{fin}}(\mathbb{C})$ for the set of all finite Dedekind $\mathbb{C}$-types which are, of course, just the finite isomorphism types, since finite functions are partial recursive.

Now we describe the language. The variables are $v_0, v_1, \ldots$. Equality is always to be interpreted by identity. The function letters are $\phi^0, \phi^1, \ldots$ and we assume a *fixed* correlation of finitary recursive strict combinatorial functors G^i with the function letters ϕ^i. The functor G^i induces, therefore, a function $\dim G^i$ on natural numbers and a function G^i on RETs. We list the interpretations below and omit sub- and superscripts for clarity.

Interpretation	Variable range	Function
Natural numbers	E	$\dim G$
Finite isomorphism types	$\Lambda^{\mathrm{fin}}(\mathbb{C})$	G restricted to $\Lambda^{\mathrm{fin}}(\mathbb{C})$
Infinite Dedekind types	$\Lambda^\infty(\mathbb{C})$	G restricted to $\Lambda'(\mathbb{C})$
Dedekind types	$\Lambda(\mathbb{C})$	G restricted to $\Lambda(\mathbb{C})$
All $\mathbb{C}$-RETs	$\Omega(\mathbb{C})$	G

where G is the function on $\Omega(\mathbb{C})$ to $\Omega(\mathbb{D})$ induced by the finitary recursive strict combinatorial functor $G: \mathbb{C} \to \mathbb{D}$. Terms $t^1, t^2, \ldots$ are defined and interpreted in the usual way. We shall generally refer to interpretations simply as their universes, e.g. $E, \Lambda(\mathbb{C})$.

From Section 15 we use the finitary recursive strict combinatorial functor

$$V: \mathbb{S} \to \mathbb{C}$$

such that if $\mathfrak{A} \in \mathbb{S}$ is finite with n elements then $V(\mathfrak{A})$ is finite with dimension n. V extends naturally to $\times^n \mathbb{S} \to \mathbb{C}_0 \times \cdots \times \mathbb{C}_{n-1}$ where

$$V(\mathfrak{A}_0, \ldots, \mathfrak{A}_{n-1}) = (V\mathfrak{A}_0, \ldots, V\mathfrak{A}_{n-1})$$

(where we write $V\mathfrak{A}_i$ for $V_i \mathfrak{A}_i$ when $V_i: \mathbb{S} \to \mathbb{C}_i$).

V induces a function V on $\Lambda(\mathbb{S})$ since V is a finitary recursive strict combinatorial functor. Moreover, V is strictly monotonically increasing and its range includes spaces of all dimensions. If $\mathbb{C}_0, \mathbb{C}_{n-1}$ have dimension and $\mathfrak{A} \in \mathbb{C}_0 \times \cdots \times \mathbb{C}_{n-1}$ we write $\dim \mathfrak{A} = (\dim \mathfrak{A}_0, \ldots, \dim \mathfrak{A}_{n-1})$.

Theorem 28.1. *Let R consist of all finite isomorphism types satisfying $t^1 = t^2$. Then $\dim R$ is the set of natural numbers satisfying $t^1 = t^2$.*

Proof. Suppose $G^i\colon \mathbb{C} \to \mathbb{D}$ is a finitary recursive strict combinatorial functor. By Theorem 16.5 there is a finitary recursive strict combinatorial functor $F^i\colon \mathbb{S} \to \mathbb{S}$ such that the following diagram commutes.

$$\begin{array}{ccc}
\Omega(\mathbb{C}) & \xrightarrow{\;G^i\;} & \Omega(\mathbb{D}) \\
{\scriptstyle V}\big\uparrow & & \big\uparrow{\scriptstyle V} \\
\Omega(\mathbb{S}) & \xrightarrow[\;F^i\;]{} & \Omega(\mathbb{S})
\end{array}$$

Now let R be the set of finite isomorphism types which satisfy $\phi^1(v) = \phi^2(v)$, then letting $\mathfrak{X}$ range only over $^0\mathbb{C}$ and $\mathfrak{A}$ over $^0\mathbb{S}$,

$$R = \{\langle \mathfrak{X} \rangle \colon G^1(\mathfrak{X}) \simeq G^2(\mathfrak{X})\}.$$

Therefore

$$\dim R = \{\dim \mathfrak{X} \colon G^1(\mathfrak{X}) \simeq G^2(\mathfrak{X})\}$$
$$= \{\dim V(\mathfrak{A}) \colon G^1\, V(\mathfrak{A}) \simeq G^2\, V(\mathfrak{A})\}$$

since for each $\mathfrak{X}$ there exists $\mathfrak{A}$ with $V(\mathfrak{A}) \simeq \mathfrak{X}$, and the G^i preserve isomorphisms. But then

$$\dim R = \{\dim V(\mathfrak{A}) \colon VF^1(\mathfrak{A}) \simeq VF^2(\mathfrak{A})\}$$
$$= \{\dim V(\mathfrak{A}) \colon \dim VF^1(\mathfrak{A}) = \dim VF^2(\mathfrak{A})\},$$
$$= \{\dim V(\mathfrak{A}) \colon \operatorname{card} F^1(\mathfrak{A}) = \operatorname{card} F^2(\mathfrak{A})\}.$$

Now

$$\operatorname{card} F^i(\mathfrak{A}) = \dim VF^i(\mathfrak{A}) = \dim G^i\, V(\mathfrak{A})$$
$$= (\dim G^i)(\dim V(\mathfrak{A}))$$
$$= (\dim G^i)(\operatorname{card} \mathfrak{A}), \quad \text{so}$$
$$\dim R = \{n \colon (\dim G^1)(n) = (\dim G^2)(n)\},$$

that is, $\dim R$ is the set of natural numbers satisfying $\phi^1(v) = \phi^2(v)$ in E.

Finally, since taking dimensions commutes with composition the result for $t^1 = t^2$ follows. $\square$

Theorem 28.2. $t^1 = t^2$ *is universally satisfied in E if, and only if, it is universally satisfied in Dedekind types if, and only if, it is universally satisfied in all RETs.*

Proof. The implications from right to left are trivial due to Theorem 28.1 above.

$\phi^1(v) = \phi^2(v)$ is universally satisfied in $\Omega(\mathbb{C})$ if, and only if, the finitary recursive strict combinatorial functors G^1, G^2 induce the same function on $\Omega(\mathbb{C})$. But if $\phi^1(v) = \phi^2(v)$ is universally satisfied in E then G^1, G^2 induce the same functions on finite isomorphism types since $\dim \mathfrak{A} =$

dim $\mathfrak{B}$ if, and only if, $\mathfrak{A} \cong \mathfrak{B}$ for finite $\mathfrak{A}$, $\mathfrak{B}$. Then by Theorem 26.1 G^1, G^2 induce the same function on all Dedekind $\mathbb{C}$-RETs. By Corollary 19.6 they induce the same function on all RETs. The conclusion for $t^1 = t^2$ follows by Theorem 16.7. $\square$

Theorem 28.3. *Suppose there exists a natural number z such that for* $x_i \in \times^n E$ *if* $x_i \geq z$ *for all* $i < n$, *then x satisfies*

$$t^1(v) = t^2(v) \& \dots \& t^{2k-1}(v) = t^{2k}(v) \to r(v) = s(v).$$

Then every $X \in \times^k \Lambda^\infty(\mathbb{C})$ *satisfies the implication in* $\Lambda^\infty(\mathbb{C})$.

Proof. Since dim $\mathfrak{A}$ is large and finite if, and only if, $\mathfrak{A}$ is large and finite the theorem follows at once from Corollary 27.4. Formally, if $k(n) = $ cardinality of a space of dimension n then k is a (one-one) strictly increasing function of n. $\square$

29. Universal Sentences II

We now show that any implication that does not fall under Theorem 28.3 in fact fails in infinite Dedekind types. It is then straightforward to extend our results to all universal sentences involving only equations.

Theorem 29.1. *Suppose that for every natural number z there is an* $x \in \times^s E$ *with all co-ordinates* $\geq z$ *such that x does not satisfy*

$$t^1(v) = t^2(v) \& \dots \& t^{2k-1}(v) = t^{2k}(v) \to r(v) = s(v)$$

in E. Then there exists a Dedekind type X with all co-ordinates infinite which makes the implication false in $\Lambda^\infty(\mathbb{C})$.

Proof. We use Theorem 18.2 as this gives the clearest short proof. (As usual we collapse the argument to the one variable case so in what follows $h(0), \dots, h(n)$ is to be regarded as an n-element sequence of s-tuples $(h_0(i), \dots, h_{s-1}(i))$ for $t = 0, \dots, n$. If $h_j \colon E \to E$ are strict $\mathbb{C}$-combinatorial functions for $j = 0, \dots, s-1$ then

$$h_j(x) = \sum_{i=0}^x c_j^i \begin{bmatrix} x \\ i \end{bmatrix}$$

where all $c_j^i \geq 0$. We write this as

$$h(x) = \sum c^i \begin{bmatrix} x \\ i \end{bmatrix},$$

that is, in vector form.)

Call a finite sequence $h(0), \dots, h(n)$ of s-tuples of natural numbers *acceptable* if there are $c^0, \dots, c^n$, which are s-tuples of natural numbers with all co-ordinates >0 such that

$$h(x) = \sum c^i \begin{bmatrix} x \\ i \end{bmatrix} \quad \text{for } x = 0, \ldots, n.$$

Observe that if $h(0), \ldots, h(n-1)$ is acceptable then $h(0), \ldots, h(n-1), l$ is acceptable provided that every co-ordinate of l exceeds every co-ordinate of

$$c^0 \begin{bmatrix} n \\ 0 \end{bmatrix} + \cdots + c^{n-1} \begin{bmatrix} n \\ n-1 \end{bmatrix}$$

since $\begin{bmatrix} n \\ n \end{bmatrix} = 1$.

Now define monotonic strictly increasing recursive strict $\mathbb{C}$-combinatorial functions $h_j : E \rightarrow E$ for $j = 0, \ldots, s-1$ as follows. Suppose $h(0), \ldots, h(n)$ have been defined. Then let $h(n+1)$ be such that $h(0), \ldots, h(n+1)$ is the first sequence (in the effective enumeration of all finite sequences from $\times^s E$) such that

(1) the implication in the theorem fails when $h(n+1)$ is assigned as value to v, and

(2) $h(0), \ldots, h(n+1)$ is acceptable.

By hypothesis there are arbitrary large assignments in E satisfying (1) and by the remark above taking $h(n+1)$ sufficiently large keeps $h(0), \ldots, h(n+1)$ acceptable.

Then assigning $h(r)$ to v makes the implication false for all r.

Now let $H_0, \ldots, H_{s-1}$ be finitary recursive strict combinatorial functors inducing $h_0, \ldots, h_{s-1}$. For ease of notation let h_i be the function letter denoting h_i and write $h(v)$ for $(h_0(v), \ldots, h_{s-1}(v))$. Then

$$t^{2i-1}(h(v)) = t^{2i}(h(v))$$

is an identity in E for $0 \le i \le k$ so by Theorem 28.2 it is an identity in $\Lambda(\mathbb{C})$. Let $(H(X))_i = H_i(X)$, $i = 0, \ldots, s-1$.

Since H_i is a finitary recursive combinatorial functor, if $X \in \Lambda^\times(\mathbb{C})$ then $H_i(X) \in \Lambda(\mathbb{C})$ by Theorem 15.4. Since in all co-ordinates h is strictly increasing, we know $H(X)$ is infinite in all co-ordinates. Further, if $H(X)$ is assigned to v then the antecedent of the implication is satisfied in $\Lambda(\mathbb{C})$.

To complete the proof it suffices to show

$$r(h(v)) = s(h(v))$$

is false for at least one $X \in \Lambda^\infty(\mathbb{C})$. Suppose this is not the case, then by Corollary 18.4 there is a finite isomorphism type Y such that for all types Z with $Z \supseteq Y$,

$$r(h(v)) = s(h(v))$$

is satisfied by Z. By Theorem 28.1, if x is a natural number $\geq \dim Y$ ($=\dim \mathfrak{Y}$ for any $\mathfrak{Y} \in Y$) then x satisfies $r(h(v)) = s(h(v))$ in E. Thus all sufficiently large natural numbers x satisfy this equation. This contradicts the construction of h which made the equation false, so the theorem is established. $\square$

Theorem 29.2. *If*

$$t^1 = t^2 \& \ldots \& t^{2k-1} = t^{2k} \to r^1 = s^1 \vee \cdots \vee r^m = s^m$$

is universally satisfied in $\Lambda^\infty(\mathbb{C})$ then there exists an i such that

$$t^1 = t^2 \& \ldots \& t^{2k-1} = t^{2k} \to r^i = s^i$$

is universally satisfied in $\Lambda^\infty(\mathbb{C})$.

Proof. Suppose the conclusion fails. Then by Theorem 28.3 for each i with $1 \leq i \leq m$ and each natural number z there is an s-tuple of integers x with all $x_j \geq z$ such that the antecedent conjunction is satisfied but $r^i = s^i$ is not (when interpreted in the natural numbers). Now we repeat the construction of Theorem 29.1 choosing $h(p)$ always to satisfy the antecedent but such that if $p \equiv i \pmod{m}$ then $h(p)$ does not satisfy $r^i = s^i$. Then (with notation as in the proof of Theorem 29.1)

$$t^1(h(v)) = t^2(h(v)) \& \ldots \& t^{2k-1}(h(v)) = t^{2k}(h(v))$$

is universally satisfied. But by Corollary 18.6 (plus the remarks succeeding it) if

$$r^1(h(v)) = s^1(h(v)) \vee \cdots \vee r^m(h(v)) = s^m(h(v))$$

is universally satisfied in $\Lambda^\infty(\mathbb{C})$, then for some i, $r^i(h(v)) = s^i(h(v))$ is universally satisfied in $\Lambda^\infty(\mathbb{C})$. The rest of the proof is now identical with that for Theorem 29.1. $\square$

Finally we establish a simple necessary and sufficient condition for the truth of universal sentences in Dedekind $\mathbb{C}$-types. First we observe that we only need consider sentences of the form in Theorem 29.2. Then the proof differs from those above only in that we sometimes have to deal with finite arguments.

Lemma 29.3. *Every quantifier free formula whose atomic formulae are equations is logically equivalent to a conjunction of formulae of the form*

$$t^1 = t^2 \& \ldots \& t^{2k-1} = t^{2k} \to r^1 = s^1 \vee \cdots \vee r^m = s^m.$$

Proof. Put the formula in conjunctive normal form then each conjunct is of the form

$$\neg t^1 = t^2 \vee \cdots \vee \neg t^{2k-1} = t^{2k} \vee r^1 = s^1 \vee \cdots \vee r^m = s^m.$$

Now use $\psi \to \chi$ is equivalent to $\neg \psi \vee \chi$ and $\neg \psi^1 \vee \neg \psi^2$ is equivalent to $\neg (\psi^1 \& \psi^2)$. $\square$

Theorem 29.4. *A necessary and sufficient condition that*
$$t^1 = t^2 \& \dots \& t^{2k-1} = t^{2k} \to r^1 = s^1 \vee \cdots \vee r^m = s^m$$
is universally satisfied in $\Lambda(\mathbb{C})$ is that, whenever some of the variables occurring are assigned natural number values, there exists a natural number z and an i with $1 \le i \le m$, such that, whenever the remaining variables are assigned natural number values $\ge z$, the resulting assignment satisfies
$$t^1 = t^2 \& \dots \& t^{2k-1} = t^{2k} \to r^i = s^i \quad \text{in } E.$$

Proof. If $F(\mathfrak{X}, \mathfrak{Y})$ is a finitary recursive strict combinatorial functor of two arguments and $\mathfrak{X}^0$ is finite then $F(\mathfrak{X}^0, \mathfrak{Y})$ is a finitary recursive strict combinatorial functor of $\mathfrak{Y}$ alone. Hence if we regard the finite arguments in the implication in the $\Lambda(\mathbb{C})$ interpretation as parameters the result follows at once from Theorem 29.2 and Theorem 28.3. $\square$

Part VIII. Sound Values

30. Soundly Based Types

In this section we assume that $\mathbb{C}' = \mathbb{C}(\mathfrak{M}')$ is suitable and has dimension and $\mathfrak{M}'$ has degree 1.

Definition 30.1. A $\mathbb{C}'$-RET X is said to be *soundly based* if there is a representative $\mathfrak{A}$ in X and a recursively enumerable independent set B such that $\mathfrak{A} \cap B$ is a basis for $\mathfrak{A}$.

If $\mathbb{V}$ is the category of vector spaces over a fixed finite field, then the condition on $\mathfrak{A}$ above is Dekker's condition for an α-space (Dekker [1969]) and an RET X is soundly based if X has at least one α-space representative. (We do not assert that X then consists entirely of α-spaces.)

Let $f': E \to |\mathfrak{M}'|$ be a recursive injection whose range is an independent set and let $V': \mathbb{S} \to \mathbb{C}'$ be the corresponding closure functor (*cf.* Section 7) which is a finitary recursive strict combinatorial functor.

Lemma 30.2. *An RET X in $\Omega(\mathbb{C}')$ is soundly based if, and only if, X is in the range of V'.*

Proof. If X is in the range of V' then there exists $V'(\mathfrak{A})$ in X. But $B = f'(E)$ is a recursively enumerable independent set and $B \cap V'(\mathfrak{A})$ is a basis for $V'(\mathfrak{A})$. So X is soundly based.

Conversely, suppose X is soundly based. Let B' be a recursively enumerable independent set and $\mathfrak{A}'$ a representative of X such that $\mathfrak{A}' \cap B'$ is a basis for $\mathfrak{A}'$. Without loss of generality we may assume B' is infinite. Let τ be a one-one partial recursive function with domain B' and codomain $B = f'(E)$. Let $\mathrm{cl}\,\tau \colon \mathrm{cl}\,B' \to \mathrm{cl}\,B$ be the elementary monomorphism extending τ given by Lemma 5.2 and Lemma 6.9. Then $\mathfrak{A}' \cap B'$ is in the domain of τ so $\mathfrak{A}'$ is in the domain of $\mathrm{cl}\,\tau$. Let $\mathfrak{A} = (\mathrm{cl}\,\tau)(\mathfrak{A}')$, then $\mathfrak{A}$ is also a representative of X. But $\mathfrak{A} \cap B \subseteq f(E)$ and $\mathfrak{A} \cap B$ is a basis for $\mathfrak{A}$ so $\mathfrak{A} = V'((f')^{-1}(\mathfrak{A} \cap B))$ and X is indeed in the range of V'. $\quad\square$

Theorem 30.3. *Let $F \colon \mathbb{C} \to \mathbb{C}'$ be a partial recursive strict combinatorial functor induced by a recursive precombinatorial operator G. Then the range of F consists entirely of soundly based RETs.*

Proof. If $B = \bigcup \{G(\mathfrak{A}'): \mathfrak{A}' \in {}^{\circ}\mathbb{C}\}$ then B is a recursively enumerable independent set and, for any $\mathfrak{A}$ in $\mathbb{C}$, $B \cap F(\mathfrak{A})$ is a basis for $F(\mathfrak{A})$. Hence the RET of $F(\mathfrak{A})$ is soundly based. $\square$

The next theorem shows that, in case all the categories involved have dimension and their underlying models have degree 1, then the truth of universal statements involving functions induced by finitary recursive strict combinatorial functors does not depend on whether all Dedekind types, or merely all soundly based Dedekind types, are considered.

Theorem 30.4. *Let ψ be a quantifier-free formula built up from function letters denoting functions induced by finitary recursive strict combinatorial functors between categories with dimension whose underlying models have degree 1. Then ψ is universally satisfied in soundly based Dedekind types if, and only if, it is universally satisfied in Dedekind types.*

Proof. Let v be the function letter denoting V. Since ψ is equivalent to a formula in conjunctive normal form we need only consider formulae of the form

(1)
$$t^1(v_0, \ldots, v_{n-1}) = t^2(v_0, \ldots, v_{n-1}) \, \& $$
$$\ldots \& \, t^{2k-1}(v_0, \ldots, v_{n-1}) = t^{2k}(v_0, \ldots, v_{n-1})$$
$$\to r^1(v_0, \ldots, v_{n-1}) = s^1(v_0, \ldots, v_{n-1}) \lor$$
$$\cdots \lor r^m(v_0, \ldots, v_{n-1}) = s^m(v_0, \ldots, v_{n-1})$$

where the t^j, s^j and r^j are terms which are compositions of function letters. The formula (1) is universally satisfied in $\Lambda(\mathbb{C})$ if, and only if, the condition on E in Theorem 29.4 is satisfied. But we also have that

(2)
$$t^1(v(v_0), \ldots, v(v_{n-1})) = t^2(v(v_0), \ldots, v(v_{n-1})) \, \& \ldots$$
$$\to r^1(v(v_0), \ldots, v(v_{n-1})) = s^1(v(v_0), \ldots, v(v_{n-1})) \lor \cdots$$

is universally satisfied if, and only if, the identical condition on E of Theorem 29.4 is satisfied, since v denotes the identity function in E. Thus (1) is universally satisfied in $\Lambda(\mathbb{C})$ if, and only if, (2) is universally satisfied in $\Lambda(\mathbb{S})$. But by Lemma 30.2 the latter is precisely the condition that (1) is universally satisfied in soundly based Dedekind types. This completes the proof. $\square$

Example 30.1. In the category $\mathbb{S}$ the projections $f_0(v_0, v_1) = v_0$ and $f_1(v_0, v_1) = v_1$ from $E \times E$ to E are both strictly $\mathbb{S}$-combinatorial. Moreover in E $v(v_0) = v(v_1) \to v_0 = v_1$ reduces to $v_0 = v_1 \to v_0 = v_1$. Hence V is one-one on Dedekind types, since $V(X) = V(Y) \to X = Y$ holds for Dedekind $\mathbb{S}$-types X, Y by Theorem 30.4. We can therefore introduce the Dedekind dimension of a soundly based Dedekind type X as the

Dedekind $\mathbb{S}$-type Y such that $X = V(Y)$. This contains Dekker's extension of dimension (Dekker [1969]) but not Hamilton's (Hamilton [1970]).

Example 30.2. In $\mathbb{S}$ all types are soundly based since every subset of E is independent.

Theorem 30.5. *In the category* $\mathbb{V}$ *of vector spaces over a fixed finite field there is a continuum of non-soundly based Dedekind types.*

Before we give the proof we note that Barbara Osofsky's theorem (Dekker [1969]) that there is a continuum of subspaces (of the infinite dimensional space) with no base extendable to a recursively enumerable independent set is a more special result than the above theorem. This is because our definition of recursive equivalence is broader and this fact also means our proof is longer.

Proof. Let $\mathfrak{M}$ be a countably infinite dimensional vector space over the fixed finite field so that $\mathbb{V} = \mathbb{C}(\mathfrak{M})$. We use the strong topology on $\mathbb{V}$ defined in Section 17. By Baire's category theorem it will suffice to show that for every independent set B of $\mathfrak{M}$ and every one-one partial function p from $\mathfrak{M}$ to $\mathfrak{M}$ the set

$$A = A(B, p) = \{\mathfrak{A} \in \mathbb{C}(\mathfrak{M}): \mathfrak{A} \subseteq \operatorname{dom} p \,\&$$

$$p|\mathfrak{A} \text{ is an isomorphism } \& \, p(\mathfrak{A}) \cap B \text{ is a basis for } p(\mathfrak{A})\}$$

is nowhere dense. For then the union of the $A(B, p)$ over all pairs (B, p) such that B is recursively enumerable and p is one-one partial recursive will be a meagre set exhausting all soundly based types.

Let $U(\mathfrak{A}, N)$ be a non-empty neighbourhood in $\mathbb{C}(\mathfrak{M})$. We must select a non-empty subneighbourhood disjoint from A.

If there exists $\mathfrak{A}' \in {}^{\circ}\mathbb{C}(\mathfrak{M})$ with $\mathfrak{A}$ in $U(\mathfrak{A}, N)$, such that p is not defined on all of $\mathfrak{A}'$, then $U(\mathfrak{A}', N)$ is the required subneighbourhood.

If there exists $\mathfrak{A}' \in {}^{\circ}\mathbb{C}(\mathfrak{M})$ with $\mathfrak{A}$ in $U(\mathfrak{A}, N)$, such that p restricted to $\mathfrak{A}'$ is not a $\mathbb{C}(\mathfrak{M})$-isomorphism between $\mathfrak{A}'$ and $p(\mathfrak{A}')$, then $U(\mathfrak{A}', N)$ is the required subneighbourhood.

If B is finite with n elements then let $\mathfrak{A}'$ be an $(n+1)$-dimensional space in $U(\mathfrak{A}, N)$, then $U(\mathfrak{A}', N)$ is the required subneighbourhood.

It remains to consider the case where B is infinite and for every $\mathfrak{A}'$ in $U(\mathfrak{A}, N)$, p is defined on $\mathfrak{A}'$ and $p|\mathfrak{A}'$ is a $\mathbb{C}(\mathfrak{M})$-isomorphism from $\mathfrak{A}'$ onto $p(\mathfrak{A}')$.

If there is no $\mathfrak{A}' \in U(\mathfrak{A}, N)$, $\mathfrak{A}' \neq \mathfrak{A}$, such that $p(\mathfrak{A}') \cap B$ is a basis for $p(\mathfrak{A}')$, then any finite $\mathfrak{A}' \in U(\mathfrak{A}, N)$ with $\mathfrak{A}' \neq \mathfrak{A}$ gives the required subneighbourhood $U(\mathfrak{A}', N)$. Otherwise we may choose an $\mathfrak{A}' \in U(\mathfrak{A}, N)$, $\mathfrak{A}' \neq \mathfrak{A}$, such that $p(\mathfrak{A}') \cap B$ is a basis for $p(\mathfrak{A}')$. Since $\mathfrak{A}' \supset \mathfrak{A}$ certainly $p(\mathfrak{A}') \supset p(\mathfrak{A})$. Since $p(\mathfrak{A}') \cap B$ is a basis for $p(\mathfrak{A}')$, we can find a finite

$B^0 \subseteq B$ such that $p(\mathfrak{A}') \cap B^0$ is a basis for a space properly containing $p(\mathfrak{A})$. But then

$$\mathfrak{A}^0 = p^{-1} \operatorname{cl}\big(p(\mathfrak{A}') \cap B^0\big)$$

is in $U(\mathfrak{A}, N)$ and in ${}^\circ\mathbb{C}(\mathfrak{M})$. Since B is independent, $p(\mathfrak{A}^0) \cap B = p(\mathfrak{A}') \cap B^0$ is a basis for $p(\mathfrak{A}^0)$.

In exactly the same way there is an $\mathfrak{A}^*$ in $U(\mathfrak{A}^0, N)$ such that $p(\mathfrak{A}^*) \cap B$ is a basis for $p(\mathfrak{A}^*)$ and $\mathfrak{A}^* \supset \mathfrak{A}^0$. (For if not, then $U(\mathfrak{A}^*, N)$ is the required subneighbourhood for any finite $\mathfrak{A}^* \supset \mathfrak{A}^0$.) Since $\mathfrak{A}^* \supset \mathfrak{A}^0$, $p(\mathfrak{A}^*) \cap B$ is a basis for $p(\mathfrak{A}^*)$ and $p(\mathfrak{A}^0) \cap B$ is a basis for $p(\mathfrak{A}^0)$, there exists $b^1 \in (p(\mathfrak{A}^*) - p(\mathfrak{A}^0)) \cap B$. Let $\mathfrak{A}^1 = \operatorname{cl}(\{p^{-1}(b^1)\} \cup \mathfrak{A}^0)$, then $\mathfrak{A}^1 \in U(\mathfrak{A}^0, N)$ and $p(\mathfrak{A}^1)$ has basis $(p(\mathfrak{A}^0) \cap B) \cup \{b^1\}$. In the same way we now choose an $\mathfrak{A}^{**}$ in $U(\mathfrak{A}^1, N)$ with $\mathfrak{A}^{**} \supset \mathfrak{A}^1$ such that $p(\mathfrak{A}^{**}) \cap B$ is a basis for $p(\mathfrak{A}^{**})$ and let $b^2 \in (p(\mathfrak{A}^{**}) - p(\mathfrak{A}^1)) \cap B$. Let $\mathfrak{A}^2 = \operatorname{cl}(\{p^{-1}(b^2)\} \cup \mathfrak{A}^1)$, then $\mathfrak{A}^2 \in U(\mathfrak{A}^1, N)$ and $p(\mathfrak{A}^2)$ has basis $(p(\mathfrak{A}^0) \cap B) \cup \{b^1, b^2\}$. Now the vector $b^1 + b^2$ is in $\operatorname{cl}\{b^1, b^2\}$ so $p^{-1}(b^1 + b^2)$ is in $\mathfrak{A}^2$. Let

$$\mathfrak{A}^3 = \operatorname{cl}\big(\mathfrak{A}^0 \cup \{p^{-1}(b^1 + b^2)\}\big)$$

then $\mathfrak{A}^3 \in U(\mathfrak{A}^0, N)$. Now $(p(\mathfrak{A}^0) \cap B) \cup \{b^1, b^2\}$ is independent, so $p^{-1}(b^1) \notin \mathfrak{A}^3$. Finally therefore $U(\mathfrak{A}^3, N \cup \{p^{-1}(b^1)\})$ is a non-empty subneighbourhood and we claim it is the desired subneighbourhood. First we observe that if $b, b' \in B$, $b \neq b'$ and $B' \subseteq B$, then $b + b' \in \operatorname{cl} B'$ implies $b, b' \in B'$. So if $\mathfrak{A}^* \in U(\mathfrak{A}^3, N \cup \{p^{-1}(b^1)\})$ we have $p^{-1}(b^1 + b^2) \in \mathfrak{A}^*$ but $p^{-1}(b^1) \notin \mathfrak{A}^*$ (or equivalently $b^1 + b^2 \in p(\mathfrak{A}^*)$ but $b^1 \notin p(\mathfrak{A}^*)$). But then $p(\mathfrak{A}^*) \cap B$ does not contain b^1 so $p(\mathfrak{A}^*) \cap B$ cannot span $p(\mathfrak{A}^*)$. So $U(\mathfrak{A}^3, N \cup \{p^{-1}(b^1)\})$ is indeed the required non-empty subneighbourhood and the proof is complete. $\square$

31. Extending Partial Functions to Soundly Based Types

In this section $\mathbb{C} = \mathbb{C}_0 \times \mathbb{C}_1$ and we assume $\mathbb{C}_0 = \mathbb{C}(\mathfrak{M}_0)$, $\mathbb{C}_1 = \mathbb{C}(\mathfrak{M}_1)$, $\mathbb{C}_1$ has dimension and $\mathfrak{M}_1$ has degree 1.

Theorem 31.1. *Let h be a partial function from finite $\mathbb{C}_0$-isomorphism types to finite $\mathbb{C}_1$-isomorphism types. Then, for every Dedekind $\mathbb{C}_0$-type X there is at most one soundly based Dedekind $\mathbb{C}_1$-type Y such that $(\mathsf{X}, \mathsf{Y}) \in \mathbb{C}(h)$.*

First we remark that if we ignore non-soundly based types in $\mathbb{C}_1$ then partial functions on 0-chain types yield partial functions on Dedekind types. (However, there may be many non-soundly based types Z with (X, Z) in $\mathbb{C}(h)$ and only one or zero soundly based Y with (X, Y) in $\mathbb{C}(h)$.)

We need some lemmata.

Lemma 31.2. *A subset Γ of the objects of $\mathbb{C}_0 \times \mathbb{C}_1$ is the set of all elements attainable from a (recursive) ^0h-frame if and only if Γ is a (recursive) frame map such that for finite $\mathfrak{A}$ from dom Γ, $h(\langle\mathfrak{A}\rangle) = \langle\Gamma(\mathfrak{A})\rangle$.*

Proof. Suppose $\Gamma = \mathscr{A}(G)$ where G is a (recursive) ^0h-frame. Then $(\mathfrak{A}, \mathfrak{B}) \in G$ implies the isomorphism type of $(\mathfrak{A}, \mathfrak{B})$ is in h. So in order to complete the proof in this direction it suffices, by Theorem 23.4, to show that Γ is a function (that is, is single-valued).

So suppose $(\mathfrak{A}, \mathfrak{B}) \in \Gamma = \mathscr{A}(G)$ and $\mathfrak{A}$ is finite. We show that $(\mathfrak{A}, \mathfrak{B}) \in G$ and that $\mathfrak{B}$ is uniquely determined. $(\mathfrak{A}, \emptyset) \in G^*$ so $C_G(\mathfrak{A}, \emptyset) \in G$ and

$$C_G(\mathfrak{A}, \emptyset) \subseteq (\mathfrak{A}, \mathfrak{B}).$$

Therefore $C_G(\mathfrak{A}, \emptyset) = (\mathfrak{A}, \mathfrak{B}^0)$ for some $\mathfrak{B}^0 \subseteq \mathfrak{B}$. On the other hand if $x \in \mathfrak{B}$ then $(\emptyset, \{x\}) \subseteq (\mathfrak{A}, \mathfrak{B})$ implies $C_G(\emptyset, \{x\}) = (\mathbb{C}^0, \mathbb{C}^1) \subseteq (\mathfrak{A}, \mathfrak{B})$. In order to show that $\mathfrak{B}^0 = \mathfrak{B}$ it therefore suffices to show that for each $x \in \mathfrak{B}$ the corresponding $\mathbb{C}^1 \subseteq \mathfrak{B}^0$. Since $(\mathbb{C}^0, \mathbb{C}^1) \in G$, $C_G(\mathbb{C}^0, \emptyset) = (\mathbb{C}^0, \mathfrak{D})$ for some $\mathfrak{D}$. Now $(\mathbb{C}^0, \emptyset) \subseteq (\mathfrak{A}, \emptyset)$ implies $C_G(\mathbb{C}^0, \emptyset) = (\mathbb{C}^0, \mathfrak{D}) \subseteq C_G(\mathfrak{A}, \emptyset) = (\mathfrak{A}, \mathfrak{B}^0)$, so $\mathfrak{D} \subseteq \mathfrak{B}^0$. But $(\mathbb{C}^0, \emptyset) \subseteq (\mathbb{C}^0, \mathbb{C}^1)$ implies $C_G(\mathbb{C}^0, \emptyset) = (\mathbb{C}^0, \mathfrak{D}) \subseteq (\mathbb{C}^0, \mathbb{C}^1) \in G$. Hence $\mathfrak{D} \subseteq \mathbb{C}^1$. But now both $(\mathbb{C}^0, \mathfrak{D})$ and $(\mathbb{C}^0, \mathbb{C}^1)$ are in G. Since h is single-valued and G is an h-frame it follows that $\mathfrak{D} = \mathbb{C}^1$. Hence $\mathbb{C}^1 = \mathfrak{D} \subseteq \mathfrak{B}^0$ as was to be shown. Finally we show that $\mathscr{A}(G)$ is a map. Suppose $(\mathfrak{A}, \mathfrak{B}^0)$, $(\mathfrak{A}, \mathfrak{B}^1) \in \mathscr{A}(G)$ and let $x \in \mathfrak{B}^0$. Then $(\emptyset, \{x\}) \subseteq (\mathfrak{A}, \mathfrak{B}^0)$ and so $(\emptyset, \{x\}) \in G^*$ and $C_G(\emptyset, \{x\}) = (\mathbb{C}^0, \mathbb{C}^1) \subseteq (\mathfrak{A}, \mathfrak{B}^0)$. Now $(\mathbb{C}^0, \mathbb{C}^1) \in G$ so $C_G(\mathbb{C}^0, \emptyset) = (\mathbb{C}^0, \mathfrak{D})$ for some $\mathfrak{D}$. But G is a map, so $\mathfrak{D} = \mathbb{C}^1$, and $(\mathbb{C}^0, \emptyset) \subseteq (\mathfrak{A}, \mathfrak{B}^1)$ so since $\mathbb{C}^0 \subseteq \mathfrak{A}$ we get $C_G(\mathbb{C}^0, \emptyset) = (\mathbb{C}^0, \mathbb{C}^1) \subseteq (\mathfrak{A}, \mathfrak{B}^1)$. But $\{x\} \subseteq \mathbb{C}^1 \subseteq \mathfrak{B}^1$. Hence $\mathfrak{B}^0 \subseteq \mathfrak{B}^1$.

Similarly $\mathfrak{B}^1 \subseteq \mathfrak{B}^0$ and so $\mathfrak{B}^0 = \mathfrak{B}^1$.

The converse direction is immediate from Theorem 23.4. □

Lemma 31.3. *Let $F: \mathbb{D} \to \mathbb{C}_1$ be a recursive frame map with $\mathbb{D} \subseteq \mathbb{C}_0$. Suppose $f_1: E \to \mathfrak{M}_1$ is a recursive injection whose range is independent and which induces the closure functor $V_1: \mathbb{S} \to \mathbb{C}_1$. Suppose that for finite $\mathfrak{A}$ in $\mathbb{D}$, $F(\mathfrak{A})$ is always in the range of V_1. For $\mathfrak{A}$ in $\mathbb{D}$ define*

$$F^e(\mathfrak{A}) = F(\mathfrak{A}) - \bigcup \{F(\mathfrak{A}'): \mathfrak{A}' \subset \mathfrak{A} \ \& \ \mathfrak{A}' \in \mathbb{D}\}$$

and $G(\mathfrak{A}) = F^e(\mathfrak{A}) \cap \operatorname{ran} f_1$. Then

1) *$\mathfrak{A} \neq \mathfrak{B}$ implies $G(\mathfrak{A}) \cap G(\mathfrak{B}) = \emptyset$ for $\mathfrak{A}, \mathfrak{B}$ finite objects of $\mathbb{D}$,*
2) *$\bigcup \{G(\mathfrak{A}): \mathfrak{A} \in \mathbb{D} \ \& \ \mathfrak{A} \text{ is finite}\}$ is independent and*
3) *For all $\mathfrak{A} \in \mathbb{D}$,*

$$F(\mathfrak{A}) = \operatorname{cl} \bigcup \{G(\mathfrak{A}'): \mathfrak{A}' \subseteq \mathfrak{A} \ \& \ \mathfrak{A}' \text{ is finite} \ \& \ \mathfrak{A}' \in \mathbb{D}\}.$$

Proof. As for Lemma 7.7 we easily show that if $\mathfrak{A}, \mathfrak{B}$ are finite and in $\mathbb{D}$ then $F^e(\mathfrak{A})$ and $F^e(\mathfrak{B})$ are disjoint and that for all $\mathfrak{A}$ in $\mathbb{D}$, $F(\mathfrak{A})$ is the union of all $F^e(\mathfrak{A}')$ with $\mathfrak{A}' \subseteq \mathfrak{A}$, $\mathfrak{A}'$ finite and in $\mathbb{D}$. Since $G(\mathfrak{A}) \subseteq F^e(\mathfrak{A})$, 1) is obvious. Since $G(\mathfrak{A}) \subseteq f_1(E)$ and $f_1(E)$ is independent, 2) is clear.

Finally for 3) we observe that for any object $\mathfrak{B}$ in $\mathbb{S}$, $V_1(\mathfrak{B})$ has *basis* $V_1(\mathfrak{B}) \cap f_1(E)$. So from the hypothesis we have that $F(\mathfrak{A}') \cap f_1(E)$ is a basis for $F(\mathfrak{A}')$ for all finite $\mathfrak{A}' \in \mathbb{D}$. Then for finite $\mathfrak{A}$ in $\mathbb{D}$, $F(\mathfrak{A})$ is the disjoint union of all $F^e(\mathfrak{A}')$ with $\mathfrak{A}' \subseteq \mathfrak{A}$ and $\mathfrak{A}'$ in $\mathbb{D}$ so $\bigcup \{G(\mathfrak{A}'): \mathfrak{A}' \subseteq \mathfrak{A} \, \& \, \mathfrak{A}' \in \mathbb{D}\}$ is a disjoint union and constitutes a basis for $F(\mathfrak{A})$. So for finite $\mathfrak{A}$ in $\mathbb{D}$,

$$F(\mathfrak{A}) = \mathrm{cl} \bigcup \{G(\mathfrak{A}'): \mathfrak{A}' \subseteq \mathfrak{A} \, \& \, \mathfrak{A}' \in \mathbb{D}\}.$$

The corresponding assertion for arbitrary $\mathfrak{A}$ in $\mathbb{D}$ follows by continuity (Lemma 22.3) since $F(\mathfrak{A})$ is the union of all $F(\mathfrak{A}')$ with $\mathfrak{A}' \subseteq \mathfrak{A}$ when $\mathfrak{A}'$ is finite and in $\mathbb{D}$. $\square$

Lemma 31.4. *Let $F^1, F^2: \mathbb{D} \to \mathbb{C}_1$ be recursive frame maps where $\mathbb{D} \subseteq \mathbb{C}_0$. Let $f_1: E \to \mathfrak{M}_1$ be a recursive injection with independent range and suppose V_1 is the induced closure functor. Suppose that for finite $\mathfrak{A}$ in $\mathbb{D}$, $F^1(\mathfrak{A})$ is $\mathbb{C}_1$-isomorphic to $F^2(\mathfrak{A})$ and that $F^1(\mathfrak{A})$, $F^2(\mathfrak{A})$ are both in the range of V_1. Then for all $\mathfrak{A} \in \mathbb{D}$,*

$$F^1(\mathfrak{A}) \simeq F^2(\mathfrak{A}).$$

Proof. Define F^{1e}, G^1 from F^1 and F^{2e}, G^2 from F^2 using Lemma 31.3. We claim that for all finite $\mathfrak{A}$ in $\mathbb{D}$, $G^1(\mathfrak{A})$ and $G^2(\mathfrak{A})$ have the same cardinality. If $\mathfrak{A}$ is the smallest element in $\mathbb{D}$, $F^1(\mathfrak{A}) \simeq F^2(\mathfrak{A})$ implies $\dim F^1(\mathfrak{A}) = \dim F^2(\mathfrak{A})$. But in this case $G^i(\mathfrak{A})$ is a basis for $F^i(\mathfrak{A})$ so card $G^1(\mathfrak{A}) = \mathrm{card}\, G^2(\mathfrak{A})$. Now suppose $\mathfrak{A}$ is finite and in $\mathbb{D}$ and for all finite $\mathfrak{A}'$ in $\mathbb{D}$ with $\mathfrak{A}' \subseteq \mathfrak{A}$ and $\mathfrak{A}' \neq \mathfrak{A}$, card $G^1(\mathfrak{A}') = \mathrm{card}\, G^2(\mathfrak{A}')$. By Lemma 31.3 for $i = 1, 2$ we have

$$\dim F^i(\mathfrak{A}) = \sum \{\mathrm{card}\, G^i(\mathfrak{A}'): \mathfrak{A}' \subseteq \mathfrak{A} \, \& \, \mathfrak{A}' \in \mathbb{D}\}.$$

By hypothesis $\dim F^1(\mathfrak{A}) = \dim F^2(\mathfrak{A})$ and by the induction hypothesis card $G^1(\mathfrak{A}') = \mathrm{card}\, G^2(\mathfrak{A}')$ for all $\mathfrak{A}' \subset \mathfrak{A}$ with $\mathfrak{A}' \in \mathbb{D}$. Hence, subtracting, card $G^1(\mathfrak{A}) = \mathrm{card}\, G^2(\mathfrak{A})$.

Now, as for Theorem 7.6, define a one-one partial recursive function τ with domain

$$\bigcup \{G^1(\mathfrak{A}'): \mathfrak{A}' \text{ is finite} \, \& \, \mathfrak{A}' \in \mathbb{D}\}$$

and codomain

$$\bigcup \{G^2(\mathfrak{A}'): \mathfrak{A}' \text{ is finite} \, \& \, \mathfrak{A}' \in \mathbb{D}\}$$

mapping each $G^1(\mathfrak{A}')$ bijectively onto $G^2(\mathfrak{A}')$ and then the restriction $\tau_{\mathfrak{A}}$ of $\mathrm{cl}\,\tau$ to $F^1(\mathfrak{A})$ is a $\mathbb{C}_1$-isomorphism of $F^1(\mathfrak{A})$ onto $F^2(\mathfrak{A})$, that is $\tau_{\mathfrak{A}}\colon F^1(\mathfrak{A})\simeq F^2(\mathfrak{A})$ for all $\mathfrak{A}\in\mathbb{D}$. $\square$

Proof of Theorem 31.1. Suppose (X,Y) and (X,Z) are in $\mathbb{C}(h)$ and Y,Z are soundly based. Let $\mathfrak{A}^1\in\mathsf{X}$ and choose $\mathfrak{A}^2\in\mathsf{Y}$, $\mathfrak{A}^3\in\mathsf{Z}$ so that $\mathfrak{A}^2,\mathfrak{A}^3$ are in the range of V_1. By Theorem 25.5 there is a recursive ^0h-frame K^1 such that $(\mathfrak{A}^1,\mathfrak{A}^2)\in\mathscr{A}(K^1)$ and a recursive ^0h-frame K^2 such that $(\mathfrak{A}^1,\mathfrak{A}^3)\in\mathscr{A}(K^2)$. Then $\mathscr{A}(K^1)=F^1$ and $\mathscr{A}(K^2)=F^2$ are frame maps by Lemma 31.2. Suppose domain F^i is $\mathbb{D}^i$ then $\mathbb{D}^1=\mathscr{A}(L^1)$ and $\mathbb{D}^2=\mathscr{A}(L^2)$ where L^1,L^2 are recursive frames. By Lemma 27.1, since $\mathfrak{A}^1$ is Dedekind and attainable from both L^1 and L^2, $\mathfrak{A}^1\in\mathscr{A}(L^1\cap L^2)$. But then by Theorem 23.5, F^1,F^2 restricted to $\mathscr{A}(L^1\cap L^2)$ are recursive frame maps and we can apply Lemma 31.4. Hence $F^1(\mathfrak{A}^1)\simeq F^2(\mathfrak{A}^1)$, that is $\mathsf{Y}=\mathsf{Z}$ as required. $\square$

32. Functions from Infinite Dedekind Types to Soundly Based Dedekind Types

We take $\mathbb{C}$, $\mathbb{C}_0$, $\mathbb{C}_1$ as in the previous section except we do not initially require $\mathfrak{M}_1$ has degree 1.

Theorem 32.1. *Let h be a partial function from finite $\mathbb{C}_0$-isomorphism types to finite $\mathbb{C}_1$-isomorphism types. Suppose that for any $\mathsf{X}\in\Lambda^\infty(\mathbb{C}_0)$ there is a soundly based $\mathsf{Y}\in\Lambda(\mathbb{C}_1)$ with $(\mathsf{X},\mathsf{Y})\in\mathbb{C}(h)$. Then there exists $\mathfrak{A}^0$ in $^\circ\mathbb{C}_0$, a recursive frame map Φ with domain $\{\mathfrak{A}\in\mathbb{C}_0\colon \mathfrak{A}^0\subseteq\mathfrak{A}\}$ and codomain $\mathbb{C}_1=\mathbb{C}(\mathfrak{M}_1)$ and a recursively enumerable independent set $B\subseteq\mathfrak{M}_1$ such that*

1) $\mathfrak{A}\in{}^\circ\mathbb{C}_0$ *and* $\mathfrak{A}\supseteq\mathfrak{A}^0$ *imply* $\langle\Phi(\mathfrak{A})\rangle=h\langle\mathfrak{A}\rangle$,

and

2) *if* $\mathfrak{A}\in{}^\circ\mathbb{C}_0$ *and* $\mathfrak{A}\supseteq\mathfrak{A}^0$ *then* $\mathrm{cl}\bigl(\Phi(\mathfrak{A})\cap B\bigr)=\Phi(\mathfrak{A})$.

If all infinite types in $\mathfrak{M}_0$ are dense, then the converse holds.

Proof. We use the fact from Lemma 31.2 that H is a ^0h-frame if and only if $\Phi=\mathscr{A}(H)$ is a frame map such that for finite $\mathfrak{A}$ in the domain of Φ, $\langle\Phi(\mathfrak{A})\rangle=h\langle\mathfrak{A}\rangle$. Let Φ be such a frame map, B an infinite independent set in $\mathfrak{M}_1$ and let

$$A(\Phi,B)=\{\mathfrak{A}\in\mathbb{C}_0\colon \Phi(\mathfrak{A})\text{ is defined and }\mathrm{cl}\bigl(\Phi(\mathfrak{A})\cap B\bigr)=\Phi(\mathfrak{A})\}.$$

We need a lemma using this notation.

Lemma 32.2. *Suppose that for any non-empty neighbourhood $U(\mathfrak{A}, N)$ of $\mathbb{C}_0$ there exists a finite $\mathfrak{A}^0$ in $U(\mathfrak{A}, N)$ such that either $\Phi(\mathfrak{A}^0)$ is undefined or $\mathrm{cl}(\Phi(\mathfrak{A}^0) \cap B) \neq \Phi(\mathfrak{A}^0)$. Then $A(\Phi, B)$ is nowhere dense.*

Proof. We produce a non-empty subneighbourhood of $U(\mathfrak{A}, N)$ disjoint from $A(\Phi, B)$.

Case 1. There is a finite $\mathfrak{A}^0$ in $U(\mathfrak{A}, N)$ such that $\Phi(\mathfrak{A}^0)$ is undefined. Now $\Phi = \mathscr{A}(\mathrm{H})$ where H is a recursive ^0h-frame. Then either $C_\mathrm{H}(\mathfrak{A}^0, \emptyset)$ is not defined or $C_\mathrm{H}(\mathfrak{A}^0, \emptyset) = (\mathfrak{A}', \mathfrak{B}')$ where $\mathfrak{A}^0 \subset \mathfrak{A}'$. In the former case or if $\mathfrak{A}' \cap N \neq \emptyset$ then $U(\mathfrak{A}^0, N)$ is the desired subneighbourhood. Otherwise since $\mathfrak{A}' - \mathfrak{A}^0 \neq \emptyset$ choose $x \in \mathfrak{A}' - \mathfrak{A}^0$ and then $U(\mathfrak{A}^0, N \cup \{x\})$ is the desired subneighbourhood.

Case 2. Case 1 fails. Then for some finite $\mathfrak{A}^0$ in $U(\mathfrak{A}, N)$, $\mathrm{cl}(\Phi(\mathfrak{A}^0) \cap B) \subset \Phi(\mathfrak{A}^0)$. Choose $x \in \Phi(\mathfrak{A}^0) - \mathrm{cl}(\Phi(\mathfrak{A}^0) \cap B)$. If $x \notin \mathrm{cl}\, B$, then for all $\mathfrak{A}$ in $U(\mathfrak{A}^0, N)$ we have $x \in \Phi(\mathfrak{A})$ and $x \notin \mathrm{cl}(\Phi(\mathfrak{A}) \cap B)$ so $U(\mathfrak{A}^0, N)$ is the desired subneighbourhood. If $x \in \mathrm{cl}\, B$ then $x \notin \mathrm{cl}(\Phi(\mathfrak{A}^0) \cap B)$ implies $\mathrm{supp}(x) \nsubseteq \Phi(\mathfrak{A}^0) \cap B$ (see Section 6). Let $b^0 \in \mathrm{supp}\, x - (\Phi(\mathfrak{A}^0) \cap B)$. If for all $\mathfrak{A} \in U(\mathfrak{A}^0, N)$, $b^0 \notin \Phi(\mathfrak{A})$ then for all such $\mathfrak{A}$ we have $\mathrm{supp}\, x \nsubseteq \Phi(\mathfrak{A}) \cap B$ and $x \in \Phi(\mathfrak{A})$ so again $U(\mathfrak{A}^0, N)$ is the desired subneighbourhood. If $b^0 \in \Phi(\mathfrak{A}^*)$ for some $\mathfrak{A}^* \in U(\mathfrak{A}^0, N)$, then let $\mathfrak{A}^1$ be the least (finite) such $\mathfrak{A}^*$. Since $b^0 \notin \Phi(\mathfrak{A}^0)$, $\mathfrak{A}^0 \subset \mathfrak{A}^1$ so there exists $a \in \mathfrak{A}^1 - \mathfrak{A}^0$ and then the desired subneighbourhood is $U(\mathfrak{A}^0, N \cup \{a\})$. This completes the proof of the lemma and we now complete the proof of the theorem. $\square$

Suppose, for a contradiction, that for each recursive Φ and recursively enumerable independent set B, $A(\Phi, B)$ is nowhere dense in $\mathbb{C}_0$. Since the finite and the non-Dedekind $\mathfrak{A}$ in $\mathbb{C}_0$ are first category, the Baire category theorem implies the existence of (continuum many) infinite Dedekind $\mathfrak{A}$ in $\mathbb{C}_0$ such that, for no $\mathfrak{B}$ in $\mathbb{C}_1$ is there a recursive ^0h-frame H and an infinite recursively enumerable independent set B such that $(\mathfrak{A}, \mathfrak{B}) \in \mathscr{A}(\mathrm{H})$ and $\mathrm{cl}(\mathfrak{B} \cap B) = \mathfrak{B}$. We claim that if $\mathsf{X} = \langle \mathfrak{A} \rangle$ there is then no soundly based Y with $(\mathsf{X}, \mathsf{Y}) \in \mathbb{C}(h)$. This will give the required contradiction. Suppose, therefore, that there were such a Y then there exist $\mathfrak{A}'$ in X and $\mathfrak{B}$ in Y with $(\mathfrak{A}', \mathfrak{B})$ attainable from some recursive ^0h-frame and $\mathrm{cl}(\mathfrak{B} \cap B) = \mathfrak{B}$ for a recursively enumerable independent set B. Since $\mathfrak{A}' \simeq \mathfrak{A}$ and $\mathfrak{B} \simeq \mathfrak{B}$, Theorem 25.5 shows that $(\mathfrak{A}, \mathfrak{B})$ is attainable from a recursive ^0h-frame H. If $\Phi = \mathscr{A}(\mathrm{H})$, $\Phi(\mathfrak{A}) = \mathfrak{B}$ and $\mathrm{cl}(\mathfrak{B} \cap B) = \mathfrak{B}$, contrary to the choice of $\mathfrak{A}$.

By Lemma 32.2 it follows that there is a non-empty subneighbourhood $U(\mathfrak{A}^0, N)$, an infinite recursively enumerable independent set B and a recursive frame map $\Phi = \mathscr{A}(\mathrm{H})$ (for some recursive ^0h-frame H) such that, for every finite $\mathfrak{A}$ in $U(\mathfrak{A}^0, N)$, $\Phi(\mathfrak{A})$ is defined and $\mathrm{cl}(\Phi(\mathfrak{A}) \cap B) = \Phi(\mathfrak{A})$. This implies by continuity the corresponding assertion for all $\mathfrak{A}$ in $U(\mathfrak{A}^0, N)$.

By Lemma 5.6 there is a recursive monomorphism p from $\mathfrak{M}_0$ into $\mathfrak{M}_0$ such that $p(\mathfrak{M}_0) \in U(\mathfrak{A}^0, N)$ and p is the identity on $\mathfrak{A}^0$. But then $F(\mathfrak{A}) = \Phi(p(\mathfrak{A}))$ is the desired recursive frame map. $\square$

As usual Theorem 32.1 also holds when $\mathbb{C}_0$ is itself a product category, or equivalently, h is a function of several arguments possibly arising from different categories.

Theorem 32.3. *Let* $F: \mathbb{C}_0 \to \mathbb{C}_1$ *be a recursive frame map such that*

i) *If* $\mathfrak{A}, \mathfrak{B} \in {}^\circ\mathbb{C}_0$ *and* $\mathfrak{A} \simeq \mathfrak{B}$ *in* $\mathbb{C}_0$ *then* $F(\mathfrak{A}) \simeq F(\mathfrak{B})$ *in* $\mathbb{C}_1$, *and*

ii) *There is a recursively enumerable independent set* $B \subseteq \mathfrak{M}_1$ *such that for all* $\mathfrak{A}$ *in* ${}^\circ\mathbb{C}_0$, $\mathrm{cl}(F(\mathfrak{A}) \cap B) = F(\mathfrak{A})$.

Then there is a recursive precombinatorial operator inducing F.

Proof. Let $F^e(\mathfrak{A}) = F(\mathfrak{A}) - \bigcup \{F(\mathfrak{A}'): \mathfrak{A}' \subset \mathfrak{A} \,\&\, \mathfrak{A}' \in {}^\circ\mathbb{C}_0\}$. Since F is a frame map, for $\mathfrak{A}$ in domain of F, $F(\mathfrak{A})$ is the disjoint union of all $F^e(\mathfrak{A}')$ with $\mathfrak{A}' \subseteq \mathfrak{A}$. Define an operator $G: {}^\circ\mathbb{C}_0 \to \mathscr{P}(\mathfrak{M}_1)$ by setting $G(\mathfrak{A}) = F^e(\mathfrak{A}) \cap B$. We first show G is precombinatorial. By the construction of F^e, $\mathfrak{A} \neq \mathfrak{B}$ implies $G(\mathfrak{A}) \cap G(\mathfrak{B}) = \emptyset$ so the requirement 1) for a precombinatorial operator is satisfied.

Now for any $\mathfrak{A}$ in ${}^\circ\mathbb{C}_0$,

$$F(\mathfrak{A}) \cap B = \bigcup \{F^e(\mathfrak{A}'): \mathfrak{A}' \subseteq \mathfrak{A} \,\&\, \mathfrak{A}' \in {}^\circ\mathbb{C}_0\} \cap B$$
$$= \bigcup \{G(\mathfrak{A}'): \mathfrak{A}' \subseteq \mathfrak{A} \,\&\, \mathfrak{A}' \in {}^\circ\mathbb{C}_0\}$$

so by ii) we have $F(\mathfrak{A}) = \mathrm{cl} \bigcup \{G(\mathfrak{A}'): \mathfrak{A}' \subseteq \mathfrak{A} \,\&\, \mathfrak{A}' \in {}^\circ\mathbb{C}_0\}$. Since F is continuous this equation holds for all $\mathfrak{A}$ in $\mathbb{C}_0$ so Definition 7.3 is satisfied. Finally we must show that if $\mathfrak{A}, \mathfrak{B} \in {}^\circ\mathbb{C}_0$ and $\mathfrak{A} \simeq \mathfrak{B}$ then card $G(\mathfrak{A}) =$ card $G(\mathfrak{B})$. We use i). The proof is by induction on the cardinality of $\mathfrak{A}$.

If this cardinality is minimal then $F(\mathfrak{A}) = \mathrm{cl}\, G(\mathfrak{A})$ and $F(\mathfrak{B}) = \mathrm{cl}\, G(\mathfrak{B})$ so dim $F(\mathfrak{A}) =$ dim $F(\mathfrak{B})$ implies card $G(\mathfrak{A}) =$ card $G(\mathfrak{B})$. Suppose the implication holds for all $\mathfrak{A} \in {}^\circ\mathbb{C}_0$ of smaller cardinality. Let $p: \mathfrak{A} \to \mathfrak{B}$ be a $\mathbb{C}_0$-isomorphism. We must prove that $F^e(\mathfrak{A}) \cap B$ and $F^e(\mathfrak{B}) \cap B$ have the same cardinality. By hypothesis

$$\mathrm{dim}\, F(\mathfrak{A}) = \sum \{\mathrm{card}(F^e(\mathfrak{A}') \cap B): \mathfrak{A}' \subseteq \mathfrak{A} \,\&\, \mathfrak{A}' \in {}^\circ\mathbb{C}_0\}$$

and

$$\mathrm{dim}\, F(\mathfrak{B}) = \sum \{\mathrm{card}(F^e(\mathfrak{B}') \cap B): \mathfrak{B}' \subseteq \mathfrak{B} \,\&\, \mathfrak{B}' \in {}^\circ\mathbb{C}_0\}.$$

But $\{\mathfrak{A}' \in {}^\circ\mathbb{C}_0: \mathfrak{A}' \subset \mathfrak{A}\}$ and $\{\mathfrak{B}' \in {}^\circ\mathbb{C}_0: \mathfrak{B}' \subset \mathfrak{B}\}$ are equinumerous and are perfectly matched by restrictions of p so that $p|\mathfrak{A}': \mathfrak{A}' \simeq \mathfrak{B}'$. By the induction hypothesis, card $(F^e(\mathfrak{A}') \cap B) =$ card $(F^e(\mathfrak{B}') \cap B)$. By condition i), dim $F(\mathfrak{A}) =$ dim $F(\mathfrak{B})$ since $\mathfrak{A} \simeq \mathfrak{B}$ hence, by subtraction,

$$\mathrm{card}(F^e(\mathfrak{A}) \cap B) = \mathrm{card}(F^e(\mathfrak{B}) \cap B).$$

Finally G is a recursive precombinatorial operator because F is a recursive frame map and B is recursively enumerable (cf. also Lemma 26.2). $\square$

Corollary 32.4. *With hypothesis as for Theorem 32.3 plus the assumption that $\mathfrak{M}_1$ has degree 1, F is the restriction to objects of a finitary recursive strict combinatorial functor.*

Proof. Apply Theorems 16.2, 7.4 and the formula for F in terms of G. $\square$

Theorem 32.1 becomes clearer if we consider it more model-theoretically. Let $\mathfrak{A}^0 \subseteq \mathfrak{M}_0$ be a finite algebraically closed set and $(\mathfrak{M}_0, \mathfrak{A}^0)$ be the model obtained by adding constants for all elements of $\mathfrak{A}^0$. If $\mathfrak{M}_0$ is suitable then $\mathfrak{A}^0$ being finite implies $(\mathfrak{M}_0, \mathfrak{A}^0)$ is also suitable (exercise). The objects of the category $\mathbb{C}(\mathfrak{M}_0, \mathfrak{A}^0)$ are the objects $\mathfrak{A}$ of $\mathbb{C}(\mathfrak{M}_0)$ which contain $\mathfrak{A}^0$ and the morphisms are those morphisms of $\mathbb{C}(\mathfrak{M}_0)$ whose domain contains $\mathfrak{A}^0$ and which are the identity on $\mathfrak{A}^0$.

By an *eventually finitary recursive strict combinatorial functor F* from $\mathbb{C}(\mathfrak{M}_0)$ to $\mathbb{C}(\mathfrak{M}_1)$ we mean an operator F such that for some finite $\mathfrak{A}^0$ in $\mathbb{C}(\mathfrak{M}_0)$, $F: \mathbb{C}(\mathfrak{M}_0, \mathfrak{A}^0) \to \mathbb{C}(\mathfrak{M}_1)$ is a finitary recursive strict combinatorial functor (in the usual sense). Combining these notions and the results of this section we get:

Corollary 32.5. *Let $\mathbb{C}_0 = \mathbb{C}(\mathfrak{M}_0)$ and $\mathbb{C}_1 = \mathbb{C}(\mathfrak{M}_1)$ where $\mathfrak{M}_1$ has dimension and degree 1. Let h be a partial function on finite isomorphism types. Suppose h induces a partial function on all of $\Lambda^\infty(\mathbb{C}(\mathfrak{M}_0))$ to soundly based types in $\Lambda(\mathbb{C}(\mathfrak{M}_1))$. Then there is an eventually finitary recursive strict combinatorial functor $F: \mathbb{C}(\mathfrak{M}_0, \mathfrak{A}^0) \to \mathbb{C}(\mathfrak{M}_1)$ such that for all $\mathfrak{A}$ in $\mathbb{C}(\mathfrak{M}_0, \mathfrak{A}^0)$ we have*

$$(\langle \mathfrak{A} \rangle, \langle F(\mathfrak{A}) \rangle) \in \mathbb{C}(h). \quad \square$$

In case $\mathbb{C}(\mathfrak{M}_0)$ has the property that all infinite types are dense (cf. Lemma 17.2) we observe that *all* infinite Dedekind types are mapped by the one F.

33. Total Functions to Soundly Based Dedekind Types

Let $\mathbb{C}(\mathfrak{M}_0), \ldots, \mathbb{C}(\mathfrak{M}_k)$ be suitable and suppose $\mathbb{C}(\mathfrak{M}')$ is suitable, $\mathfrak{M}'$ has dimension and $\mathfrak{M}'$ has degree 1. Let $\mathbb{C} = \mathbb{C}(\mathfrak{M}_0) \times \cdots \times \mathbb{C}(\mathfrak{M}_k)$ and $\mathbb{C}' = \mathbb{C}(\mathfrak{M}')$. If h is a partial function from finite $\mathbb{C}$-isomorphism types to finite $\mathbb{C}'$-isomorphism types, then by the results of Section 31 h yields a partial function, which we denote by h_Λ, from $\Lambda(\mathbb{C})$ to soundly based types in $\Lambda(\mathbb{C}')$ given by $h_\Lambda(X_0, \ldots, X_k) = Y$ if, and only if, Y is the (unique) soundly based Dedekind $\mathbb{C}'$-type such that $(X_0, \ldots, X_k, Y) \in (\mathbb{C} \times \mathbb{C}')(h)$. Since finite isomorphism types are the same as the finite RETs we may regard h_A as an extension of h and write $h \subseteq h_A$.

We now ask, for which functions h is h_A a total function? That is, for which h is dom $h_A = \Lambda(\mathbb{C})$? From Section 31 we have that if $F: \mathbb{C} \to \mathbb{C}'$ is a recursive frame map and h is a partial function on finite isomorphism types such that $F \cap (^\circ\mathbb{C} \times {}^\circ\mathbb{C}')$ is a ${}^\circ h$-frame, then whenever $\mathfrak{A}$ is Dedekind and in dom F and $F(\mathfrak{A})$ is soundly based then

$$h_A\langle\mathfrak{A}\rangle = \langle F(\mathfrak{A})\rangle.$$

In particular, if F is a finitary recursive strict combinatorial functor then $F = h_A$ where h is the restriction of F to finite isomorphism types. So there is a large supply of functions h with h_A total.

A function h on finite isomorphism types is said to be *eventually recursive strict combinatorial* if there exist $\mathfrak{A}_0 \in {}^\circ\mathbb{C}(\mathfrak{M}_0), \ldots, \mathfrak{A}_k \in {}^\circ\mathbb{C}(\mathfrak{M}_k)$ and a finitary recursive strict combinatorial functor

$$F: \mathbb{C}(\mathfrak{M}_0, \mathfrak{A}_0) \times \cdots \times \mathbb{C}(\mathfrak{M}_k, \mathfrak{A}_k') \to \mathbb{C}'$$

such that for all $\mathfrak{B}$ with $\mathfrak{B}_i \in {}^\circ\mathbb{C}(\mathfrak{M}_i, \mathfrak{A}_i)$ and $\mathfrak{B}_i \supseteq \mathfrak{A}_i$ for all $i \leq k$ we have

$$h(\langle\mathfrak{B}_0\rangle, \ldots, \langle\mathfrak{B}_k\rangle) = \langle F(\mathfrak{B}_0, \ldots, \mathfrak{B}_k)\rangle.$$

Theorem 33.1. *Suppose all infinite types are dense in* $\mathbb{C}(\mathfrak{M}_0), \ldots, \mathbb{C}(\mathfrak{M}_k)$. *Then* h_A *is total if, and only if, every function obtained from* h *by specifying finite isomorphism types as parameters at some but not all arguments of* h *is eventually recursive strict combinatorial.*

Proof. Let h_1 be any function obtained from h by specifying parameters. Then the proof of Theorem 32.1 for such h_1 yields that if h_A is total then h satisfies the condition of the theorem. Conversely, suppose the condition holds. We must show h_A is defined on every

$$X = (X_0, \ldots, X_k) \in \Lambda(\mathbb{C}(\mathfrak{M}_0)) \times \cdots \times \Lambda(\mathbb{C}(\mathfrak{M}_k)).$$

Some of the X_i may be finite. For simplicity of notation suppose $k = 2$, X_0 is finite and X_1, X_2 are not. By hypothesis there exist $\mathfrak{A}_1 \in {}^\circ\mathbb{C}(\mathfrak{M}_1)$ and $\mathfrak{A}_2 \in {}^\circ\mathbb{C}(\mathfrak{M}_2)$ and a finitary recursive strict combinatorial functor $F: \mathbb{C}(\mathfrak{M}_1, \mathfrak{A}_1) \times \mathbb{C}(\mathfrak{M}_2, \mathfrak{A}_2) \to \mathbb{C}(\mathfrak{M}')$ such that for all $\mathfrak{B}_1 \in {}^\circ\mathbb{C}(\mathfrak{M}_1, \mathfrak{A}_1)$ and $\mathfrak{B}_2 \in {}^\circ\mathbb{C}(\mathfrak{M}_2, \mathfrak{A}_2)$ we have

$$h(X_0, \langle\mathfrak{B}_1\rangle, \langle\mathfrak{B}_2\rangle) = \langle F(\mathfrak{B}_1, \mathfrak{B}_2)\rangle.$$

Let $\mathfrak{A}_0$ be a representative of X_0. Then

$$K = \{(\mathfrak{A}_0, \mathfrak{B}_1, \mathfrak{B}_2, F(\mathfrak{B}_1, \mathfrak{B}_2)): \mathfrak{B}_1 \in {}^\circ\mathbb{C}(\mathfrak{M}_1, \mathfrak{A}_1) \& \mathfrak{B}_2 \in {}^\circ\mathbb{C}(\mathfrak{M}_2, \mathfrak{A}_2)\}$$

is a recursive ^0h-frame and

$$\mathscr{A}(K)=\{(\mathfrak{A}_0, \mathfrak{B}_1, \mathfrak{B}_2, F(\mathfrak{B}_1, \mathfrak{B}_2)): \mathfrak{B}_1 \in \mathbb{C}(\mathfrak{M}_1, \mathfrak{A}_1) \& \mathfrak{B}_2 \in \mathbb{C}(\mathfrak{M}_2, \mathfrak{A}_2)\}.$$

By hypothesis X_1 is dense in $\mathfrak{M}_1$ so there is a representative $\mathfrak{B}_1$ of X_1 with $\mathfrak{B}_1 \supseteq \mathfrak{A}_1$ and similarly there exists $\mathfrak{B}_2$ in X_2 with $\mathfrak{B}_2 \supseteq \mathfrak{A}_2$. Then $h_A(\mathsf{X}_0, \mathsf{X}_1, \mathsf{X}_2) = \langle F(\mathfrak{B}_1, \mathfrak{B}_2) \rangle.$ □

A function satisfying the condition of Theorem 33.1 will be said to be *almost recursive strict combinatorial.*

Theorem 32.1 and Theorem 23.6 yield the following corollary to Theorem 33.1.

Corollary 33.2. *Suppose all models considered have dimension, degree 1, and all infinite types are dense. Then almost recursive strict combinatorial functions are closed under composition and identities on finite isomorphism types yield identities on $\Lambda(\mathbb{C})$.*   □

Similar theorems may be obtained for the function dim h on dimension induced by h (*cf.* Section 28).

Part IX. The Automorphism Extension Property

34. The Automorphism Extension Property

In some categories which arise from fully effective $\aleph_0$-categorical models (in particular linear orderings) we do not have dimension but do have a weaker property which we now discuss. However, there are still other categories (Boolean algebras, for example) where this fails.

Definition 34.1. A category $\mathbb{C}$ is said to have the *automorphism extension property* if whenever $\mathfrak{A}, \mathfrak{B} \in {}^\circ\mathbb{C}$ and $\mathfrak{A} \subseteq \mathfrak{B}$ then any automorphism of $\mathfrak{A}$ can be extended to an automorphism of $\mathfrak{B}$.

Example 34.1. $\mathbb{S}, \mathbb{L}, \mathbb{V}$ have this property. For $\mathbb{S}$ it is trivial since if $\mathfrak{A}, \mathfrak{B} \in {}^\circ\mathbb{C}$ then map $\mathfrak{B} - \mathfrak{A}$ one-one, onto itself to extend any automorphism of $\mathfrak{A}$.

In $\mathbb{L}$ since there is a unique automorphism of any finite linearly ordered set, namely the identity, the property trivially holds for $\mathbb{L}$.

In $\mathbb{V}$ if $\mathfrak{A} \subseteq \mathfrak{B}$ then there is a complement $\mathbb{C}$ such that $\mathfrak{A} \oplus \mathbb{C} = \mathfrak{B}$. So the identity on $\mathbb{C}$ with an automorphism of $\mathfrak{A}$ induces an automorphism of $\mathfrak{B}$. This argument extends to the general dimension case but we leave the proof of this, the next lemma, to the reader.

Lemma 34.2. *Let $\mathbb{C}$ be a category with dimension then $\mathbb{C}$ has the automorphism extension property.* $\square$

Example 34.2. $\mathbb{B}$ does not have the automorphism extension property. For let $\mathbb{B}$ be an eight element Boolean algebra with three atoms b, b', b'' and let $\mathfrak{A}$ be the subalgebra generated by $b, b' \vee b''$. Now the automorphism of $\mathfrak{A}$ given by $b \leftrightarrow b' \vee b''$ cannot be extended to an automorphism of $\mathfrak{B}$ since $b' \vee b''$ is not an atom of $\mathfrak{B}$ while b is such an atom.

The following lemmata give some useful characterizations of the automorphism extension property.

Lemma 34.3. *$\mathbb{C}$ has the automorphism extension property if, and only if, for any finite morphism p with domain $\mathfrak{A}$ and codomain $\mathfrak{B}$ and any automorphism t of $\mathfrak{A}$ there is an automorphism t' of $\mathfrak{B}$ such that the follow-*

ing diagram commutes:

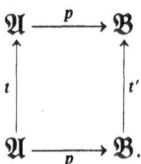

Proof. Clearly the condition is sufficient. Conversely, given p and t the map $p(a) \to p\,t(a)$ defines an automorphism of $p(\mathfrak{A})$. By the automorphism extension property this map extends to an automorphism $t': \mathfrak{B} \to \mathfrak{B}$. But then $t'\,p(a) = p\,t(a)$ for $a \in A$ hence the diagram commutes as required. □

The next lemma says roughly that if *some* finite isomorphism can be finitely extended then *any* finite isomorphism can be.

Lemma 34.4. $\mathbb{C}$ *has the automorphism extension property if, and only if, given finite morphisms* $p: \mathfrak{A} \to \mathfrak{B}$ *and* $p': \mathfrak{A}' \to \mathfrak{B}'$ *and (finite) isomorphisms* $t: \mathfrak{A} \cong \mathfrak{A}'$ *and* $t': \mathfrak{B} \cong \mathfrak{B}'$ *such that the following diagram commutes:*

$$
\begin{array}{ccc}
\mathfrak{A}' & \xrightarrow{\;p'\;} & \mathfrak{B}' \\
\Big\uparrow{\scriptstyle t} & & \Big\uparrow{\scriptstyle t'} \\
\mathfrak{A} & \xrightarrow{\;p\;} & \mathfrak{B}
\end{array}
$$

then for any isomorphism $r: \mathfrak{A} \cong \mathfrak{A}'$ *there is an isomorphism* $r': \mathfrak{B} \cong \mathfrak{B}'$ *such that*

$$
\begin{array}{ccc}
\mathfrak{A}' & \xrightarrow{\;p'\;} & \mathfrak{B}' \\
\Big\uparrow{\scriptstyle r} & & \Big\uparrow{\scriptstyle r'} \\
\mathfrak{A} & \xrightarrow{\;p\;} & \mathfrak{B}
\end{array}
$$

commutes.

Proof. Suppose $\mathbb{C}$ has the automorphism extension property. Consider the diagram below.

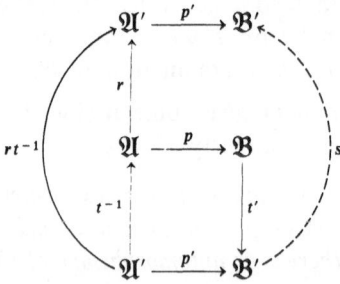

$r t^{-1}$ is an automorphism of $\mathfrak{A}'$ so by Lemma 34.3 the dotted morphism s exists and makes the outer square commute. But then putting $r' = s t'$ we complete the top half to a commutative square

$$\mathfrak{A}' \xrightarrow{\ p'\ } \mathfrak{B}'$$

$$r \qquad\qquad st' = r'$$

$$\mathfrak{A} \xrightarrow{\ p\ } \mathfrak{B}$$

as required.

The converse is obvious.   □

35. Regressive Types and Tree Frames

We shall use the automorphism extension property and the properties of regressive types established in this section to extend our earlier results to categories without dimension.

Definition 35.1. A set A of natural numbers is said to be *regressive* if there is an enumeration of A (not necessarily effective) without repetitions $a^0, a^1, a^2, \ldots$ and a partial recursive function r such that $r(a^n) = a^{n-1}$ for $n > 0$ and $r(a^0) = a^0$. r is called a *regressing function* for A.

It follows that if p is a one-one recursive function with domain containing A and $p(A) = B$ then A regressive implies B regressive. For $p r p^{-1}$ is a regressing function for B when r is a regressing function for A.

Observe that every finite set is trivially regressive.

Definition 35.2. A $\mathbb{C}$-type A is said to be *regressive* if, for some (or equivalently all) $\mathfrak{A} \in$ A, $|\mathfrak{A}|$ is a regressive set.

We shall give another characterization of regressive types.

Definition 35.3. A frame F is said to be a *tree* frame if $\{\mathfrak{B}: \mathfrak{B} \in F\ \&\ \mathfrak{B} \subseteq \mathfrak{A}\}$ is linearly ordered for each $\mathfrak{A} \in F$.

So a tree frame is a frame with no diamond

$$\mathfrak{A}$$
$$\nearrow \qquad \nwarrow$$
$$\mathfrak{B} \qquad\qquad \mathfrak{C}$$
$$\nwarrow \qquad \nearrow$$
$$\mathfrak{D}$$

in it. Note that any branch, that is, maximal linearly ordered (under $\subseteq$) subset of F is, in fact, well-ordered.

Definition 35.4. A chain $\mathfrak{A}^0 \subseteq \cdots \subseteq \mathfrak{A}^n \subseteq \cdots$ of elements of a frame F is said to be a *refinement* of a chain $\mathfrak{B}^0 \subseteq \cdots \subseteq \mathfrak{B}^m \subseteq \cdots$ if every $\mathfrak{B}^i$ is an $\mathfrak{A}^j$; the chain is said to be a *composition chain* if i) all the $\mathfrak{A}^j$ are distinct and ii) the chain has no proper refinement with all elements distinct.

Lemma 35.5. *If* F *is a tree frame and* $\mathfrak{A} \in$ F *then there is a unique composition chain with last element* $\mathfrak{A}$. *If* F *is a recursive tree frame then the composition chain is uniformly explicitly computable from an explicit index for* $\mathfrak{A}$.

Proof. If $\mathfrak{A} \in$ F, $\mathfrak{A}$ is finite. Compute $C_F(B)$ for all subsets of $|\mathfrak{A}|$. If F is a tree frame the $C_F(B)$ are linearly ordered by $\subseteq$. Removing repetitions then clearly gives a composition chain for $\mathfrak{A}$. Since any $\mathfrak{B} \in$ F with $\mathfrak{B} \subseteq \mathfrak{A}$ is of the form $C_F(B)$ for some $B \subseteq |\mathfrak{B}|$ the chain is unique. Finally we note that if F is recursive $C_F(B)$ computes an explicit index from an explicit index of B so the second part of the lemma follows. $\square$

Corollary 35.6. *If* F *is a tree frame,* $\mathfrak{A} \in \mathscr{A}(F)$ *and* $\mathfrak{A}$ *is infinite then there is a unique composition chain* $\mathfrak{A}^0 \subseteq \mathfrak{A}^1 \subseteq \cdots$ *of type* ω *such that each* $\mathfrak{A}^i \in$ F *and* $\mathfrak{A} = \bigcup \mathfrak{A}^i$.

Proof. Let $\mathfrak{B}^0 \subset \mathfrak{B}^1 \subset \cdots$ be a chain of elements of F, without repetitions, such that $\bigcup \mathfrak{B}^i = \mathfrak{A}$. Such exists since $\mathfrak{A} \in \mathscr{A}(F)$ and is infinite. Since the composition chains for the $\mathfrak{B}^i$ are unique the composition chain for $\mathfrak{B}^i$ is an initial segment of the composition chain for $\mathfrak{B}^{i+1}$ and all together constitute a composition chain for $\mathfrak{A}$. The uniqueness of the composition chain for each $\mathfrak{B}^i$ implies the uniqueness of that for $\mathfrak{A}$. $\square$

For $\mathfrak{A} \in \mathscr{A}(F)$, define
$$C_F^e \mathfrak{A} = \mathfrak{A} - \bigcup \{\mathfrak{B} \in F : \mathfrak{B} \subset \mathfrak{A}\}.$$

Suppose $\mathfrak{A} \in$ F and $\mathfrak{A}$ is not the minimal object in F, then $C_F^e(\mathfrak{A}) = \mathfrak{A} - \mathfrak{A}^-$ where $\mathfrak{A}^-$ is the element immediately preceding $\mathfrak{A}$ in the composition chain for $\mathfrak{A}$; and $C_F^e(\mathfrak{A}) = \emptyset$ for $\mathfrak{A}$ infinite. Note that when F is recursive if $\mathfrak{A}$ is finite, $\ulcorner C_F^e(\mathfrak{A}) \urcorner$ is effectively computable from $\ulcorner \mathfrak{A} \urcorner$ by Lemma 35.5. (First find the composition chain, then its penultimate element.)

Theorem 35.7. *A* $\mathbb{C}$-*type* A *is regressive if, and only if, some (every)* $\mathfrak{A}$ *with* $\langle \mathfrak{A} \rangle = $ A *is attainable from some recursive tree frame* F.

Proof. The assertion holds for finite A by the remark preceding Definition 35.2 and the fact that if $\mathfrak{A}$ is finite, $\{\mathfrak{A}\}$ is a recursive tree frame from which $\mathfrak{A}$ is attainable. So assume $\mathfrak{A}$ is infinite and attainable from a recursive tree frame F.

For any $\mathfrak{B}, \mathfrak{C} \in F$ with $\mathfrak{B} \neq \mathfrak{C}$ we have $C_F^e(\mathfrak{B}) \cap C_F^e(\mathfrak{C}) = \emptyset$ by the same proof as for Lemma 7.7(i). Let $\mathfrak{A}^0 \subset \mathfrak{A}^1 \subset \cdots$ be the composition chain for $\mathfrak{A}$ then $\{C_F^e(\mathfrak{A}^i): i = 0, 1, \ldots\}$ forms a partition of $\mathfrak{A}$. Since the elements of $C_F^e(\mathfrak{A}^i)$ are natural numbers we can enumerate $\mathfrak{A}$ as follows: enumerate $C_F^e(\mathfrak{A}^i)$ in order of magnitude after all elements of all $C_F^e(\mathfrak{A}^j)$ with $j < i$ have been enumerated. The regressing function r is defined as follows. Compute $\mathfrak{B} = C_F\{a\}$ then compute the (finite) composition chain $\mathfrak{B}^0 \subset \cdots \subset \mathfrak{B}^m (= \mathfrak{B})$ for $\mathfrak{B}$. Since $\mathfrak{B} = C_F\{a\}$, we conclude that $a \in C_F^e(\mathfrak{B}^m)$. If a is not the least element in $C_F^e(\mathfrak{B}^m)$ let $r(a)$ be the next smaller element of $C_F^e(\mathfrak{B}^m)$. If a is the least element let $r(a)$ be the largest element of $C_F^e(\mathfrak{B}^{m-1})$ which exists if $m \neq 0$. If $m = 0$ and a is the least element of $C_F^e(\mathfrak{B}^0)$ set $r(a) = a$.

Clearly r is partial recursive and $r(a^n) = a^{n-1}$ if $n > 0$, $r(a^0) = a^0$ for $\{a^i: i = 0, 1, \ldots\}$ the enumeration of $\mathfrak{A}$ given above.

If $\mathfrak{A}$ is infinite and regressive but not Dedekind then $\mathfrak{A}$ is recursively enumerable (as follows easily from Dekker and Myhill [1960], p. 112). In this case if $a^0, a^1, \ldots$ is an effective enumeration of $\mathfrak{A}$ then $\{\mathrm{cl}\{a^0, \ldots, a^n\}: n = 0, 1, 2, \ldots\}$ is clearly a recursive tree frame from which $\mathfrak{A}$ is attainable. Now suppose that $\mathfrak{A} \in \mathfrak{C}$ is infinite, Dedekind and regressive with an enumeration $a^0, a^1, \ldots$ without repetitions and regressing function r. A number b in $\mathfrak{M}$ will be called *acceptable* if $b, r(b), \ldots, r^k(b)$ are all defined and distinct and $r^k(b) = a^0$. k is said to be the *height* of b. Let F be the set of all finite objects $\mathfrak{B} \in \mathfrak{C}$ such that $\mathfrak{B} = \{b, r(b), r^2(b), \ldots, r^k(b)\}$ for some acceptable b of height k. We claim that F is a recursive frame from which $\mathfrak{A}$ is attainable.

F is a frame for suppose $\mathfrak{B}, \mathfrak{C} \in F$ and $\mathfrak{B} = \{b, r(b), \ldots, r^k(b)\}$ and $\mathfrak{C} = \{c, r(c), \ldots, r^l(c)\}$ where $\mathfrak{B} \neq \emptyset \neq \mathfrak{C}$. Now $r^k(b) = a^0 = r^l(c)$ so $\mathfrak{B} \cap \mathfrak{C} \neq \emptyset$. Let d be the element of $\mathfrak{B} \cap \mathfrak{C}$ of largest height, m say. Then $\mathfrak{B} \cap \mathfrak{C} = \{d, r(d), \ldots, r^m(d)\}$ and $r^m(d) = a^0$. But then $\mathfrak{B} \cap \mathfrak{C} \in F$. (Note that $\mathfrak{B} \cap \mathfrak{C}$ is algebraically closed since $\mathfrak{B}, \mathfrak{C}$ are.)

F is a tree frame, for, if $\mathfrak{B} = \{b, r(b), \ldots, r^k(b)\}$ and $\mathfrak{C} \subseteq \mathfrak{B}$ where $\mathfrak{C} \in F$ then $\mathfrak{C} = \{r^l(b), r^{l+1}(b), \ldots, r^k(b)\}$ for some $l \leq k$. Hence the $\subseteq$-predecessors of $\mathfrak{B}$ are linearly ordered.

We claim F is a recursive frame. First, it is clear that F is recursively enumerable since r is partial recursive. Let a finite $B \in F^*$ be given, then $C_F(B)$ is the smallest object $\mathfrak{C}$ such that (i)–(iii) below hold:

    (i) $B \subseteq |\mathfrak{C}|$,

    (ii) $|\mathfrak{C}|$ is of the form $\{c, r(c), \ldots, r^k(c)\}$ where $r^k(c) = a^0$,

    (iii) $\mathfrak{C}$ is algebraically closed.

Since taking the algebraic closure is effective and r is partial recursive we can effectively compute $C_F(B)$. So F is indeed a recursive frame.

Finally $\mathfrak{A}$ is attainable from F. For suppose B is finite and contained in $\mathfrak{A}$. Since $\mathfrak{A}$ satisfies (i), (iii) and is closed under r, there is a smallest object $\mathfrak{C} \subseteq \mathfrak{A}$ satisfying (i), (iii), and closed under r. Since $\mathfrak{A}$ is Dedekind and the procedure for finding $\mathfrak{C}$ generates a recursively enumerable subset of $\mathfrak{A}$, $\mathfrak{C}$ is finite. So $\mathfrak{C} \in F$ and $B \subseteq \mathfrak{C} \subseteq \mathfrak{A}$ and hence $\mathfrak{A}$ is attainable from F.   □

Corollary 35.8. *Let R be a set of ω-chain types. Suppose X is regressive and $X \in \mathbb{C}(^1R)$. Then if $\mathfrak{A} \in X$ there is a recursive ^1R-frame F, which is a tree frame, such that $\mathfrak{A}$ is attainable from F.*

Proof. By Theorem 25.5, $\mathfrak{A} \in \mathscr{A}(G)$ for some recursive ^1R-frame G. By Theorem 35.7, $\mathfrak{A} \in \mathscr{A}(H)$ for some recursive tree frame H. By Lemma 27.1, $F = G \cap H$ is a recursive frame from which $\mathfrak{A}$ is attainable. Now $F \subseteq G$ implies F is a ^1R-frame and $F \subseteq H$ implies F is a tree frame.   □

If $\mathbb{C}$ is a category of the form $\mathbb{C}_0 \times \cdots \times \mathbb{C}_{k-1}$ then a *regressive* object $\mathfrak{A}$ in $\mathbb{C}$ is a k-tuple of objects $(\mathfrak{A}_0, \ldots, \mathfrak{A}_{k-1})$ with $\mathfrak{A}_i \in \mathbb{C}_i$ such that

(i) there exists an enumeration without repetitions $a^0, a^1, a^2, \ldots$ where each a^i is a k-tuple $(a^i_0, \ldots, a^i_{k-1})$ such that $a^0_j, a^1_j, a^2_j, \ldots$ is an enumeration of $\mathfrak{A}_j$ and

(ii) there is a partial recursive function r from a subset of $\times^k E$ into $\times^k E$ such that

$$r(a^n_0, \ldots, a^n_{k-1}) = (a^{n-1}_0, \ldots, a^{n-1}_{k-1}) \text{ for } n > 0$$

and

$$r(a^0_0, \ldots, a^0_{k-1}) = (a^0_0, \ldots, a^0_{k-1}).$$

This notion of a regressive k-tuple is not the same as that of a k-tuple of regressive objects. This is an immediate consequence of the next theorem, for Dekker [1965] has shown that there exist regressive $\mathbb{S}$-RETs whose sum is not regressive whilst Theorem 35.10 below shows that the sum of a regressive pair (A_1, A_2) is again regressive. We call a k-tuple $A = (A_0, \ldots, A_{k-1})$ *regressive* if there exist $\mathfrak{A}_i \in A_i$ such that $(\mathfrak{A}_0, \ldots, \mathfrak{A}_{k-1})$ (considered as an object of $\mathbb{C}$) is regressive. By Theorem 35.7, a k-tuple of Dedekind types $A = (A_0, \ldots, A_{k-1})$ is regressive if, and only if, for all $\mathfrak{A}_i \in A_i$, $(\mathfrak{A}_0, \ldots, \mathfrak{A}_{k-1})$ is attainable from a recursive tree frame.

Lemma 35.9. *Let $F: \mathbb{C} \to \mathbb{C}'$ (where $\mathbb{C} = \mathbb{C}_0 \times \cdots \times \mathbb{C}_{k-1}$) be a finitary recursive combinatorial functor. Let $G \subseteq {}^0\mathbb{C}$ be a recursive tree frame and $H = \{F(\mathfrak{B}): \mathfrak{B} \in G\}$ then H is a recursive tree frame. If $\mathfrak{A}$ is attainable from G then $F(\mathfrak{A})$ is attainable from H.*

Proof. If $\mathfrak{B}^1, \mathfrak{B}^2 \in G$ then $\mathfrak{B}^1 \cap \mathfrak{B}^2 \in G$ since G is a frame. Since F preserves intersection of objects $F(\mathfrak{B}^1) \cap F(\mathfrak{B}^2) = F(\mathfrak{B}^1 \cap \mathfrak{B}^2)$ so H is closed under intersections. Since F is finitary H is a frame.

Since G is recursively enumerable and F is finitary recursive, $\ulcorner H \urcorner$ (and therefore $\ulcorner H^* \urcorner$) is recursively enumerable.

Next we have to show that for $B \in H^*$ the map $\ulcorner C_H \urcorner : \ulcorner B \urcorner \rightsquigarrow \ulcorner C_H(B) \urcorner$ is partial recursive. By the definition of H, $C_H(B) = F(\mathbb{C})$ for some $\mathbb{C} \in G$. Now F is a combinatorial functor so $B \subseteq F(\mathbb{C})$ if, and only if, for all $x \in B$, $F^-(x) \subseteq \mathbb{C}$ if, and only if, $\bigcup \{F^-(x): x \in B\} \subseteq \mathbb{C}$. Then

$$C_H(B) = \bigcap \{F(\mathbb{C}): B \subseteq F(\mathbb{C}) \,\&\, \mathbb{C} \in G\}$$
$$= F(\bigcap \{\mathbb{C}: B \subseteq F(\mathbb{C}) \,\&\, \mathbb{C} \in G\})$$

(since F preserves intersections)

$$= F(\bigcap \{\mathbb{C}: \bigcup \{F^-(x): x \in B\} \subseteq \mathbb{C} \,\&\, \mathbb{C} \in G\})$$
$$= F(C_G(\bigcup \{F^-(x): x \in B\})).$$

But F is finitary recursive, and $\ulcorner C_G \urcorner$ and $\ulcorner F \urcorner$ are partial recursive functions so $C_H(B)$ is effectively computable as required.

In order to show that H is a tree frame it suffices to show that if $\mathfrak{B}' \in H$, $\mathfrak{D} \in G$ and $\mathfrak{B}' \subseteq F(\mathfrak{D})$ then there exists $\mathfrak{B} \in G$ with $\mathfrak{B} \subseteq \mathfrak{D}$ and $\mathfrak{B}' = F(\mathfrak{B})$. For then if $\mathfrak{B}', \mathfrak{C}', \mathfrak{D}' \in H$ with $\mathfrak{B}', \mathfrak{C}' \subseteq \mathfrak{D}'$ there exist $\mathfrak{B}, \mathfrak{C}, \mathfrak{D} \in G$ with $\mathfrak{D}' = F(\mathfrak{D})$ and $\mathfrak{B}' = F(\mathfrak{B})$, $\mathfrak{C}' = F(\mathfrak{C})$ but $\mathfrak{B}, \mathfrak{C} \subseteq \mathfrak{D}$. Since G is a tree frame either $\mathfrak{B} \subseteq \mathfrak{C}$ or $\mathfrak{C} \subseteq \mathfrak{B}$. Since F preserves inclusions $F(\mathfrak{B}) \subseteq F(\mathfrak{C})$ or $F(\mathfrak{C}) \subseteq F(\mathfrak{B})$ so H is then a tree frame.

Assume therefore that $\mathfrak{B}' \in H$ and $\mathfrak{B}' \subseteq F(\mathfrak{D})$ for some $\mathfrak{D} \in G$. Certainly there exists $\mathfrak{B}^0$ such that $\mathfrak{B}' = F(\mathfrak{B}^0)$ for some $\mathfrak{B}^0 \in G$ by the definition of H. If $\mathfrak{B}' \subseteq F(\mathfrak{D})$ then $F(\mathfrak{B}^0) = \mathfrak{B}' = \mathfrak{B}' \cap F(\mathfrak{D}) = F(\mathfrak{B}^0) \cap F(\mathfrak{D}) = F(\mathfrak{B}^0 \cap \mathfrak{D})$. But G is a frame so $\mathfrak{B}^0 \cap \mathfrak{D} \in G$, but trivially $\mathfrak{B}^0 \cap \mathfrak{D} \subseteq \mathfrak{D}$, so take $\mathfrak{B} = \mathfrak{B}^0 \cap \mathfrak{D}$, then $\mathfrak{B}$ has the required properties and H is a tree frame.

Finally, if $\mathfrak{A}$ is attainable from G then $F(\mathfrak{A})$ is attainable from H. For if B is finite then $B \subseteq F(\mathfrak{A})$ implies $C_G(\bigcup \{F^-(x): x \in B\}) \subseteq \mathfrak{A}$ as we showed above. Now $\bigcup \{F^-(x): x \in B\}$ is finite (since F is combinatorial and B is finite) and contained in $\mathfrak{A}$ so $\mathbb{C} = C_G(\bigcup \{F^-(x): x \in B\})$ is finite and contained in $\mathfrak{A}$. But $\mathfrak{A}$ is attainable from G so $\mathbb{C} \in G$; but then $F(\mathbb{C}) \in H$ and is finite as F is finitary. Therefore $B \subseteq F(\mathbb{C}) \subseteq F(\mathfrak{A})$ and the proof is complete. $\square$

Theorem 35.10. *Suppose $F: \mathbb{C} \to \mathbb{C}'$ is a finitary recursive combinatorial functor and $\mathbb{C} = \mathbb{C}_0 \times \cdots \times \mathbb{C}_{k-1}$. If $A = (A_0, \ldots, A_{k-1})$ is a regressive k-tuple in $\Lambda(\mathbb{C}_0) \times \cdots \times \Lambda(\mathbb{C}_{k-1})$ then $F(A)$ is a regressive type in $\Lambda(\mathbb{C}')$.*

Proof. By Theorem 35.7 and the remarks following Corollary 35.8 there exist $\mathfrak{A}_i \in A_i$ and a recursive tree frame G such that $\mathfrak{A} = (\mathfrak{A}_0, \ldots, \mathfrak{A}_{k-1})$ is attainable from G. By Lemma 35.9, $F(\mathfrak{A})$ is attainable from a recursive tree frame. Then $F(\mathfrak{A})$ is Dedekind by Theorem 15.4. Theorem 35.7 shows that $F(\mathfrak{A})$ is regressive, so $\mathsf{F}(\mathsf{A}) = \langle F(\mathfrak{A}) \rangle$ is regressive. $\quad\square$

36. Solutions of Equations and the Automorphism Extension Property

We now show that if G, H are finitary recursive combinatorial functors into a category with the automorphism extension property then the extension to regressive Dedekind types of the finite 1-chain type solutions to $\mathsf{G}(\mathsf{X}) = \mathsf{H}(\mathsf{X})$ is the set of regressive solutions of $\mathsf{G}(\mathsf{X}) = \mathsf{H}(\mathsf{X})$. So, in particular, the regressive solutions of $\mathsf{G}(\mathsf{X}) = \mathsf{H}(\mathsf{X})$ are entirely determined by the finite 1-chain type solutions.

Theorem 36.1. *Let* $\mathbb{C} = \mathbb{C}_0 \times \cdots \times \mathbb{C}_{k-1}$ *and suppose* $\mathbb{C}'$ *has the automorphism extension property. Let* $G, H: \mathbb{C} \to \mathbb{C}'$ *be finitary recursive combinatorial functors. Let* ^1R *be the set (for* $\mathbb{C}$*) of all finite 1-chain type solutions* X *to* $\mathsf{G}(\mathsf{X}) = \mathsf{H}(\mathsf{X})$. *Then the extension of* ^1R *to regressive Dedekind* $\mathbb{C}$*-types is the set of regressive Dedekind* $\mathbb{C}$*-type solutions* X *of* $\mathsf{G}(\mathsf{X}) = \mathsf{H}(\mathsf{X})$.

Proof. By Corollary 25.6 if X is a Dedekind $\mathbb{C}$-type solution of $\mathsf{G}(\mathsf{X}) = \mathsf{H}(\mathsf{X})$ then $\mathsf{X} \in \mathbb{C}(^1R)$.

Now suppose that F is a recursive ^1R-tree frame from which $\mathfrak{A} = \langle \mathfrak{A}_0, \dots, \mathfrak{A}_{k-1} \rangle$ is attainable where $\mathfrak{A}$ is Dedekind and $\langle \mathfrak{A} \rangle = \mathsf{X}$. Let $\mathbb{C}^0, \mathbb{C}^1, \dots$ be an effective enumeration without repetitions of F such that if $\mathbb{C} \in F$ and $\mathbb{C} \subseteq \mathbb{C}^j$ then $\mathbb{C} = \mathbb{C}^i$ for some $i \leq j$. (Such an enumeration exists since there are only a finite number of objects contained in a given finite object and if $\mathfrak{B} \subseteq \mathbb{C} \in F$ then $\mathfrak{B} \in F$ if, and only if, $C_F \mathfrak{B} = \mathfrak{B}$.) We shall recursively define a sequence of $\mathbb{C}'$-morphisms p^i such that $p^i \colon G(\mathbb{C}^i) \cong H(\mathbb{C}^i)$ and p^i extends p^j whenever $\mathbb{C}^j \subseteq \mathbb{C}^i$.

Since $\mathbb{C}^0 \in F$ the inclusion morphism $\mathbb{C}^0 \subseteq \mathbb{C}^0$ has its 1-chain type in ^1R so there is an isomorphism p^0 such that

$$
\begin{array}{ccc}
H(\mathbb{C}^0) & \subseteq & H(\mathbb{C}^0) \\
\uparrow & & \uparrow \\
{\scriptstyle p^0} & & {\scriptstyle p^0} \\
G(\mathbb{C}^0) & \subseteq & G(\mathbb{C}^0)
\end{array}
$$

commutes.

Now suppose $p^0, \dots, p^n$ have been defined. Since $\mathbb{C}^{n+1}$ is finite and in a tree frame, there is a largest $\mathfrak{B} \in F$ with $\mathfrak{B} \subset \mathbb{C}^{n+1}$. By the choice of the enumeration $\mathfrak{B} = \mathbb{C}^j$ for some $j \leq n$. Let $i \colon \mathbb{C}^j \to \mathbb{C}^{n+1}$ be the inclusion map. Since $\mathbb{C}^j, \mathbb{C}^{n+1} \in F$, the 1-chain type $\langle i \rangle$ of i is in ^1R so $\mathsf{G}\langle i \rangle = \mathsf{H}\langle i \rangle$

therefore there exist isomorphisms r, s such that

$$H(\mathbb{C}^j) \xrightarrow{\;H(i)\;} H(\mathbb{C}^{n+1})$$

$$\uparrow r \qquad\qquad \uparrow s$$

$$G(\mathbb{C}^j) \xrightarrow{\;G(i)\;} G(\mathbb{C}^{n+1})$$

commutes.

Now $p^j\colon G(\mathbb{C}^j) \cong H(\mathbb{C}^j)$ so by Lemma 34.4 (since $\mathbb{C}'$ has the automorphism extension property) there is an isomorphism p' such that

$$H(\mathbb{C}^j) \xrightarrow{\;H(i)\;} H(\mathbb{C}^{n+1})$$

$$\uparrow p^j \qquad\qquad \uparrow p'$$

$$G(\mathbb{C}^j) \xrightarrow{\;G(i)\;} G(\mathbb{C}^{n+1})$$

commutes. Now there is a uniform effective enumeration of all finite isomorphisms in $\mathbb{C}'$ so let p^{n+1} be the first p' in this enumeration which makes the above diagram commute. Hence by induction we can define p^n for all n. Moreover, since being an extension is transitive, if $\mathbb{C}^j \subseteq \mathbb{C}^n$, p^n is an extension of p^j.

Let $p = \bigcup \{p^n\colon n = 0, 1, 2, \ldots\}$. Then we claim that p is a one-one partial recursive function. p is a function for suppose $p^m(x) = y$ and $p^n(x) = y'$ then $x \in G(\mathbb{C}^m) \cap G(\mathbb{C}^n) = G(\mathbb{C}^m \cap \mathbb{C}^n)$. But $\mathbb{C}^m \cap \mathbb{C}^n = \mathbb{C}^j$ for some $j \leq m, n$ so $y = p^m(x) = p^j(x) = p^n(y) = y'$ since p^m, p^n extend p^j. Similarly $p^{-1} = \bigcup \{(p^n)^{-1}\colon n = 0, 1, 2, \ldots\}$ is a function so p is one-one. Clearly p is partial recursive since we have given a uniform method of computing the p^n.

Finally we show that if $\mathfrak{A}$ is attainable from F then p restricted to $G(\mathfrak{A})$ is a $\mathbb{C}'$-isomorphism of $G(\mathfrak{A})$ onto $H(\mathfrak{A})$. Since $\mathfrak{A}$ is attainable from F, it follows that there exists a sequence $\mathbb{C}^{i_n}$ of elements of F such that $m \leq n$ implies $\mathbb{C}^{i_m} \subseteq \mathbb{C}^{i_n}$ (and $i_m \leq i_n$), $p^{i_n}\colon G(\mathbb{C}^{i_n}) \cong H(\mathbb{C}^{i_n})$ in $\mathbb{C}'$ and $\mathfrak{A} = \bigcup \{\mathbb{C}^{i_n}\colon n = 0, 1, 2, \ldots\}$. But then, since G, H are continuous and p^{i_n} extends p^{i_m} for $i_m \leq i_n$ by construction, $p^+ = \bigcup \{p^{i_n}\colon n = 0, 1, 2, \ldots\}$ is a $\mathbb{C}'$-isomorphism of $G(\bigcup \mathbb{C}^{i_n})$ onto $H(\bigcup \mathbb{C}^{i_n})$ that is, p restricted to $\mathfrak{A}$ is a $\mathbb{C}'$-isomorphism of $G(\mathfrak{A})$ onto $H(\mathfrak{A})$ as required. $\square$

In exactly the same way as we obtained Theorem 27.3 (except that ^1R replaces ^0R and Theorem 26.1 is replaced by Theorem 36.1) we immediately get the next theorem.

Let $t^1, \ldots, t^{2k}$ be terms such that for each i, t^{2i-1}, t^{2i} both denote functions induced by (compositions of) finitary recursive combinatorial functors from $\mathbb{C}$ into the same category $\mathbb{E}^i$ and r, s denote functions induced by finitary recursive combinatorial functors from $\mathbb{C}$ into $\mathbb{D}$. Suppose that $\mathbb{D}$ has the automorphism extension property.

Theorem 36.2. *With the above hypothesis, if*

$$t^1 = t^2 \,\&\, \ldots \,\&\, t^{2k-1} = t^{2k} \to r = s$$

is true for all 1-chain types X *from* $^0\mathbb{C}$ *then it is true for all regressive Dedekind $\mathbb{C}$-types* X. $\square$

Part X. Satisfiability

37. Finitary Relations

In this last part our aim is a strong simultaneous satisfiability criterion for relations on chain types. We make no special assumption on the categories involved other than the basic one that they arise from fully effective $\aleph_0$-categorical models as suitable categories.

We shall be dealing with infinite collections of relations; the number of arguments, though finite in each case, may be unbounded. It is therefore convenient to consider a weak infinite product category and to identify n-ary relations with relations determined by the first n factors in the product.

Definition 37.1. Let $\mathbb{C}_i$ for $i=0, 1, 2, \ldots$ be suitable categories. Let $\mathfrak{O}_i$ be the minimal object in $\mathbb{C}_i$. The *weak product category* of the $\mathbb{C}_i$ has as objects sequences $\mathfrak{A}=(\mathfrak{A}_0, \mathfrak{A}_1, \ldots)$ such that $\mathfrak{A}_i$ is an object of $\mathbb{C}_i$ where, except for a finite number of i, $\mathfrak{A}_i=\mathfrak{O}_i$. The morphisms are infinite sequences $p=(p_0, p_1, \ldots)$ such that p_i is a morphism of $\mathbb{C}_i$ where, except for a finite number of i, p_i is the identity morphism on $\mathfrak{O}_i$. We denote this category by $\mathbb{C}$, its objects by $\mathfrak{A}, \mathfrak{A}^1, \mathfrak{B}^2$, etc. and its morphisms by p, p^1, q^2, etc. Composition of morphisms is defined co-ordinatewise.

Thus p_j^i is the j-th co-ordinate of the morphism p^i and p_j^i is a morphism in $\mathbb{C}_j$. It is clear that $\mathbb{C}$ is indeed a category and that there is a natural injection of each finite product $\mathbb{C}_{i_1} \times \cdots \times \mathbb{C}_{i_n}$ into $\mathbb{C}$.

$$°\mathbb{C} = \{\mathfrak{A} \in \mathbb{C}: \mathfrak{A}_i \in °\mathbb{C}_i \text{ for all } i\}.$$

We shall call the objects in $°\mathbb{C}$ the *finite* objects of $\mathbb{C}$. (They have only finite co-ordinates.)

If a is any infinite sequence (of objects, morphisms, numbers, or whatever) and n is a natural number then $a|n$ is the finite sequence

$$a|n=(a_0, a_1, \ldots, a_{n-1}),$$

of length n. If S is any set of infinite sequences, then

$$S|n=\{a|n: a\in S\}.$$

Then $\mathbb{C}|n = \mathbb{C}_0 \times \cdots \times \mathbb{C}_{n-1}$ is the category whose objects are the $\mathfrak{A}|n$ such that $\mathfrak{A}$ is an object in $\mathbb{C}$ and whose morphisms are the $p|n$ such that p is a morphism of $\mathbb{C}$. All the notions we require extend naturally to $\mathbb{C}$. In particular, an m-chain type K from $\mathbb{C}$ is an infinite sequence of m-chain types, one from each $\mathbb{C}_i$. So $\mathsf{K}|n$ is a chain type of $\mathbb{C}|n$.

If

$$(1) \qquad \mathfrak{A}^0 \xrightarrow{\;p^0\;} \mathfrak{A}^1 \xrightarrow{\;p^1\;} \cdots \xrightarrow{\;p^{n-1}\;} \mathfrak{A}^n \xrightarrow{\;p^n\;} \cdots$$

is a chain then a subchain is obtained by taking a subsequence of objects with the morphisms induced by composition. Formally, we define the morphism $p^{ij}: \mathfrak{A}^i \to \mathfrak{A}^j$ as the composition $p^{j-1} \ldots p^{i+1} p^i$. If K is the chain type of (1) then a *subchain type* L of K is the chain type of a chain

$$\mathfrak{A}^{i_0} \xrightarrow{\;g^0\;} \mathfrak{A}^{i_1} \xrightarrow{\;g^1\;} \cdots$$

where $g^k = p^{i_k i_{k+1}}$ and $i_0 < i_1 < \cdots$. L is an *initial subchain type* if L is the chain type of a chain $\mathfrak{A}^0 \xrightarrow{\;p^0\;} \mathfrak{A}^1 \xrightarrow{\;p^1\;} \cdots$ of length $\leq$ length of the chain (1), that is, L is the chain type of an initial segment of (1). L is a *final subchain type* if L is the type of a chain of the form

$$\mathfrak{A}^n \xrightarrow{\;p^n\;} \mathfrak{A}^{n+1} \xrightarrow{\;p^{n+1}\;} \cdots,$$

that is, L is the chain type of a final segment of (1).

We shall treat only those relations R on chain types which are closed under the formation of subchain types. In particular we note that the set R of all finite chain types of inclusion chains $\mathfrak{A}^0 \subseteq \mathfrak{A}^1 \subseteq \cdots \subseteq \mathfrak{A}^n$ in a frame F has this property.

Definition 37.2. A set $\mathbf{R}$ of finite ω-chains in $\mathbb{C}$ is said to be a *recursively enumerable finitary relation* if (1) $\mathbf{R}$ is recursively enumerable, (2) $\mathbf{R}$ is closed under the formation of subchains, (3) $\mathbf{R}$ is closed under isomorphism and (4) there is a natural number n, called a *support* for $\mathbf{R}$, such that for all finite ω-chains K in $\mathbb{C}$

$$K \in \mathbf{R} \quad \text{if and only if} \quad K|n \in \mathbf{R}|n.$$

(We recall that a finite ω-chain is a chain $\mathfrak{A}^0 \to \mathfrak{A}^1 \to \cdots$ in which each $\mathfrak{A}^n$ is finite, that is to say, is in $°\mathbb{C}$.)

We observe that $\mathbf{R}$ is closed under the formation of subchains if and only if $\mathbf{R}|n$ is so closed where n is a support for $\mathbf{R}$. Further, if n is a support for a recursively enumerable finitary relation $\mathbf{R}$ and $m > n$ then m is a support for $\mathbf{R}$.

A set R of finite ω-chain types is said to be *recursively enumerable finitary* if R is the set of chain types in some recursively enumerable finitary relation $\mathbf{R}$ of finite ω-chains.

The usual coding ensures that if $\mathbf{R}$ is recursively enumerable finitary and R is the set of chain types in $\mathbf{R}$ then the set of codes of chain types in R is indeed a recursively enumerable set and conversely, if R is recursively enumerable finitary, then the set of all chains with chain types in R, when coded, is recursively enumerable. Thus there is no abuse of language in passing between $\mathbf{R}$ and R.

The class of all recursively enumerable finitary relations on chain types forms a lattice which we denote by $\mathscr{L} = \mathscr{L}(\mathbb{C})$. We let $\Lambda(\mathbb{C})$ consist of all infinite sequences X with $\mathsf{X}_i \in \Lambda(\mathbb{C}_i)$.

Finally, in this section we give our principal definition which is the natural transmission to $\mathbb{C}$ of the extension procedure of Section 25. We shall identify a frame F in $\mathbb{C}_0 \times \cdots \times \mathbb{C}_{n-1}$ with the frame

$$\{\mathfrak{A} \in \mathbb{C}: \ \mathfrak{A}|n \in F \ \& \ \mathfrak{A}_i = \mathfrak{O}_i \text{ for all } i \geq n\}$$

so the notion of recursive frame is unaffected. Further, in the next section we shall use a frame in $\mathbb{C}$ rather than frames in $\mathbb{C}|n$ for some n.

Definition 37.3. Let $R \in \mathscr{L}(\mathbb{C})$. Then $\mathbb{C}(R)$ is the set of all those $\mathsf{X} \in \Lambda(\mathbb{C})$ such that if n is a support for R then

$$\mathsf{X}|n \in \mathbb{C}\left({}^{\omega}(R|n)\right).$$

We note that $\mathsf{X}|n \in \mathbb{C}\left({}^{\omega}(R|n)\right)$ means that there exist $\mathfrak{X} = (\mathfrak{X}_0, \ldots, \mathfrak{X}_{n-1})$ such that $\mathsf{X}|n = (\langle \mathfrak{X}_0 \rangle, \ldots, \langle \mathfrak{X}_{n-1} \rangle)$ and $\mathfrak{X}$ is Dedekind and attainable from some recursive frame $F \subseteq \mathbb{C}_0 \times \cdots \times \mathbb{C}_{n-1}$ such that whenever $\mathfrak{B}^0 \subseteq \mathfrak{B}^1 \subseteq \cdots \subseteq \mathfrak{B}^k$ is an inclusion chain of objects in F then the type of this chain is in $R|n$. (By Lemma 25.4 "there exists $\mathfrak{X}$" above may be replaced by "for all $\mathfrak{X}$".)

38. The Master Frame

In this section we develop a master tree frame whose nodes represent all finite chain types and which bifurcates at every possible node.

We recall that we write $\mathfrak{A} \subset \mathfrak{B}$ for $\mathfrak{A} \subseteq \mathfrak{B} \ \& \ \mathfrak{A} \neq \mathfrak{B}$. In the case that $\mathfrak{A}, \mathfrak{B} \in \mathbb{C}$, the weak infinite product, this means that $\mathfrak{A}, \mathfrak{B}$ differ in at least one co-ordinate. A chain $\mathfrak{A}^0 \subseteq \mathfrak{A}^1 \subseteq \cdots$ is said to be *strictly increasing* if $\mathfrak{A}^n \subset \mathfrak{A}^{n+1}$ for all n. We use the same terminology for chain types.

Definition 38.1. An *index* σ is a finite sequence $\sigma(0), \ldots, \sigma(k-1)$ such that each $\sigma(i)$ is of the form

(1) $\qquad\qquad\qquad\qquad (\mathfrak{A}^i, b^i)$

where (i) $\mathfrak{A}^0$ is the minimal object of $\mathbb{C}$ (that is $\mathfrak{A}_n^0 = \mathfrak{O}_n$ for all n) and $b^0 = 0$, (ii) $\mathfrak{A}^0 \subset \mathfrak{A}^1 \subset \cdots \subset \mathfrak{A}^{k-1}$ is a strictly increasing chain of finite objects of $\mathbb{C}$ and $b^1, \ldots, b^{k-1} \in \{0, 1\}$.

k is called the length of σ and we write $\mathrm{lh}(\sigma) = k$. We sometimes write $(\mathfrak{A}^i(\sigma), b^i(\sigma))$ for (1).

If σ^1, σ^2 are indices we write $\sigma^1 \leq \sigma^2$ if σ^1 is an initial segment of σ^2. Clearly $\leq$ is a partial ordering of indices and in this ordering any two indices have a greatest lower bound, written $\sigma^1 \wedge \sigma^2$, which is the largest common initial segment of σ^1 and σ^2.

According to our earlier convention $\sigma | i$ is the sequence $\sigma(0), \ldots, \sigma(i-1)$ if $i < \mathrm{lh}(\sigma)$.

Theorem 38.2. *There is a fully effective one-one assignment of a finite object $\mathfrak{A}(\sigma)$ to each index σ where $\sigma = (\mathfrak{A}^0, b^0), \ldots, (\mathfrak{A}^{k-1}, b^{k-1})$ such that*

   (i) $\mathfrak{A}(\sigma|1) = \mathfrak{A}^0$,

   (ii) $\mathfrak{A}(\sigma^1) \cap \mathfrak{A}(\sigma^2) = \mathfrak{A}(\sigma^1 \wedge \sigma^2)$, *for all indices σ^1, σ^2,*

   (iii) *if $\mathrm{lh}(\sigma) = k$, then*

$$\mathfrak{A}(\sigma|1) \subseteq \cdots \subseteq \mathfrak{A}(\sigma|k)$$

has the same chain type as

$$\mathfrak{A}^0 \subset \cdots \subset \mathfrak{A}^{k-1}.$$

Proof. We can effectively enumerate all finite objects in $\mathbb{C}$ as $\mathbb{C}^0, \mathbb{C}^1, \ldots$, say, where no object is listed until all its subobjects have been enumerated. Then we can effectively enumerate all indices in such a way that (i) all (finite) proper subsequences in σ are enumerated before σ, (ii) all indices involving at most $\mathbb{C}^0, \ldots, \mathbb{C}^{n-1}$ are enumerated before any index involving $\mathbb{C}^n$ and (iii) if two indices (of the same length) differ only in their b's then the one with $b^i = 0$ for the least i where they differ is enumerated before the one with $b^i = 1$. Write $\sigma \prec \tau$ if σ is enumerated before τ in this listing.

Now there is only one index of length 1 namely $(\mathfrak{O}, 0)$ so set $\mathfrak{A}(\mathfrak{O}, 0) = \mathfrak{O}$. Then suppose $\mathfrak{A}(\tau)$ has been defined for all $\tau \prec \sigma$. Then σ is of the form $\tau, (\mathfrak{A}^k, b^k)$ where $\tau = (\mathfrak{A}^0, b^0), \ldots, (\mathfrak{A}^{k-1}, b^{k-1})$. We use the Duplication Lemma 5.5 to choose, uniformly, an extension $\mathfrak{A}(\sigma)$ of $\mathfrak{A}(\tau)$ such that $\mathfrak{A}(\tau) \subset \mathfrak{A}(\sigma)$ and $\mathfrak{A}^{k-1} \subset \mathfrak{A}^k$ have the same chain type and

(2) $$\mathfrak{A}(\sigma) \cap \mathrm{cl} \bigcup \{\mathfrak{A}(\rho): \rho \prec \sigma\} = \mathfrak{A}(\tau).$$

This is possible since $\mathrm{cl} \bigcup \{\mathfrak{A}(\rho): \rho \prec \sigma\}$ is finite and uniformly computable from σ.

An easy induction then shows that $\mathfrak{A}(\sigma^1) \cap \mathfrak{A}(\sigma^2) = \mathfrak{A}(\sigma^1 \wedge \sigma^2)$ for all indices σ^1, σ^2. Finally, the assignment is one-one for if $\sigma^1 \neq \sigma^2$ then

$\sigma^1 \wedge \sigma^2 \neq \sigma^1$ or $\neq \sigma^2$. Suppose the former holds, then if we had $\mathfrak{A}(\sigma^1) = \mathfrak{A}(\sigma^2)$ we should have $\mathfrak{A}(\sigma^1) = \mathfrak{A}(\sigma^1) \cap \mathfrak{A}(\sigma^2) = \mathfrak{A}(\sigma^1 \wedge \sigma^2)$ where $\sigma^1 \wedge \sigma^2$ is a proper initial segment of σ^1. So it suffices to show that if σ^* is a proper extension of σ then $\mathfrak{A}(\sigma) \neq \mathfrak{A}(\sigma^*)$ but this follows immediately from (2). $\square$

Theorem 38.3 (The Master Frame).

(i) $\mathrm{MF} = \{\mathfrak{A}(\sigma): \sigma \text{ is an index}\}$ *is a recursive tree frame in* $\mathbb{C}$.

(ii) $\mathrm{MF}|n = \{\mathfrak{A}(\sigma)|n: \sigma \text{ is an index}\}$ *is a recursive tree frame in* $\mathbb{C}|n$.

Proof. (i) Recall that by a finite object in $\mathbb{C}$ we mean an object with all co-ordinates finite, that is, $\mathfrak{A} \in \mathbb{C}$ is finite if, for all i, $\mathfrak{A}_i \in {}^\circ\mathbb{C}_i$. If $\mathfrak{A}(\sigma), \mathfrak{A}(\tau) \in \mathrm{MF}$ then $\mathfrak{A}(\sigma) \cap \mathfrak{A}(\tau) = \mathfrak{A}(\sigma \wedge \tau)$ by Theorem 38.2(ii) and $\sigma \wedge \tau$ is an index so $\mathfrak{A}(\sigma) \cap \mathfrak{A}(\tau) \in \mathrm{MF}$. By Theorem 38.2 MF is recursively enumerable. Suppose $B \in \mathrm{MF}^*$ and is finite then $B \subseteq \mathfrak{A}(\sigma)$ for some index σ which can be effectively computed from B since the correspondence $\sigma \rightsquigarrow \mathfrak{A}(\sigma)$ is fully effective. Let σ^0 be such an index of shortest length then $C_{\mathrm{MF}}(B) = \mathfrak{A}(\sigma^0)$. Because of the way the $\mathfrak{A}(\tau)$ are defined (see the first part of proof of Theorem 38.2) we can effectively find σ^0 from any σ such that $B \subseteq \mathfrak{A}(\sigma)$. Hence $\ulcorner C_{\mathrm{MF}} \urcorner$ is partial recursive and MF is recursive. Finally, MF is a tree frame for suppose $\mathfrak{A}, \mathfrak{B}, \mathbb{C} \in \mathrm{MF}$ where $\mathfrak{A}, \mathfrak{B} \subseteq \mathbb{C}$. Then if $\mathbb{C} = \mathfrak{A}(\sigma|r)$ we have $\mathfrak{A} = \mathfrak{A}(\sigma|m)$ and $\mathfrak{B} = \mathfrak{A}(\sigma|n)$ for some $m, n \leq r$. But then $\mathfrak{A} \subseteq \mathfrak{B}$ or $\mathfrak{B} \subseteq \mathfrak{A}$.

Part (ii) follows at once by restricting attention to the first n co-ordinates. We call MF the *master frame*. $\square$

For the next result we need the strong product category $\mathbb{C}^\#$. Its objects are all infinite sequences $\mathfrak{A}$ of objects $\mathfrak{A}_i$ in $\mathbb{C}_i$. Its morphisms are all infinite sequences p of morphisms p_i in $\mathbb{C}_i$. Composition is co-ordinatewise. We put ${}^\circ\mathbb{C}^\# = {}^\circ\mathbb{C}$, so ${}^\circ\mathbb{C} = {}^\circ\mathbb{C}^\# \subseteq \mathbb{C} \subseteq \mathbb{C}^\#$.

Definition 38.4. An *infinite index* $\bar{\sigma}$ is an infinite sequence every initial segment of which is an index.

Thus given an infinite index $\bar{\sigma}$ then we set

$$\mathfrak{A}(\bar{\sigma}) = \mathfrak{A}(\bar{\sigma}|1) \cup \mathfrak{A}(\bar{\sigma}|2) \cup \cdots.$$

Since $\bar{\sigma}|m$ is an initial segment of $\bar{\sigma}|n$ for $m \leq n$, we get $\mathfrak{A}(\bar{\sigma}|m) \subseteq \mathfrak{A}(\bar{\sigma}|n)$ for $m \leq n$, so $\mathfrak{A}(\bar{\sigma})$ is an object of $\mathbb{C}^\#$.

Corollary 38.5. *The map* $\bar{\sigma} \rightsquigarrow \mathfrak{A}(\bar{\sigma})$ *is a bijection from infinite indices to the non-finite objects attainable from the master frame* MF *in* $\mathbb{C}^\#$.

Proof. Since $\mathfrak{A}(\bar{\sigma}) = \bigcup \{\mathfrak{A}(\bar{\sigma}|n): n = 1, 2, \ldots\}$ we have $\mathfrak{A}(\bar{\sigma}) \in \mathscr{A}(\mathrm{MF})$. Conversely, if $\mathfrak{B} \in \mathscr{A}(\mathrm{MF})$ we can find an infinite sequence of finite objects in MF, $\mathfrak{A}(\sigma^0) \subseteq \mathfrak{A}(\sigma^1) \subseteq \cdots$ whose union is $\mathfrak{B}$. But $\mathfrak{A}(\sigma) \subseteq \mathfrak{A}(\tau)$

implies $\sigma \subseteq \tau$. So the σ^i are compatible (that is $\sigma^i = \sigma^j|\mathrm{lh}(\sigma^i)$ for $i \leq j$) so let $\bar{\sigma}$ be the infinite index such that $\bar{\sigma}|\mathrm{lh}(\sigma^i) = \sigma^i$.

Suppose $\bar{\sigma}, \bar{\tau}$ are infinite indices. If $\bar{\sigma} \neq \bar{\tau}$ then there exists n such that $\bar{\sigma}|n = \bar{\tau}|n$ but $\bar{\sigma}|n+1 \neq \bar{\tau}|n+1$. But then for all $m > n$, $\mathfrak{A}(\bar{\sigma}|m) \cap \mathfrak{A}(\bar{\tau}|m) = \mathfrak{A}(\bar{\sigma}|n)$ while $\mathfrak{A}(\bar{\sigma}|m) \supset \mathfrak{A}(\bar{\sigma}|n)$. Hence $\mathfrak{A}(\bar{\sigma}) \neq \mathfrak{A}(\bar{\tau})$. Thus the map $\bar{\sigma} \rightsquigarrow \mathfrak{A}(\bar{\sigma})$ is one-one and the corollary is proved. $\square$

In the next section we shall see how the master frame yields elements of $\mathbb{C}(R)$ for $R \in \mathscr{L}(\mathbb{C})$.

39. Satisfiability

Definition 39.1. An infinite sequence $\mathfrak{A}^0 \subseteq \mathfrak{A}^1 \subseteq \cdots$ of finite *objects* of $\mathbb{C}$ (or their chain type) is said to be *eventually in R* if there exists m such that for all $n \geq m$ the chain type of $\mathfrak{A}^m \subseteq \cdots \subseteq \mathfrak{A}^n$ is in R; otherwise, it is said to be *frequently out* of R.

In this section we shall prove (Theorem 39.5 and its corollary) that for each strictly increasing chain K of objects of the category $^\circ\mathbb{C}$ there is an infinite sequence $\mathbf{Z}$ of (regressive) Dedekind types such that for all R, $\mathbf{Z} \in \mathbb{C}(R)$ if, and only if, the chain type of K is eventually in R.

Lemma 39.2. *Let $\bar{\sigma}$ be the infinite index* $(\mathfrak{A}^0, b^0)$, $(\mathfrak{A}^1, b^1)$, ... *and let R be a recursively enumerable finitary relation on ω-chain types. Suppose the chain type of* $\mathfrak{A}^0 \subseteq \mathfrak{A}^1 \subseteq \cdots$ *is eventually in R and $\mathfrak{A}(\bar{\sigma})$ is Dedekind, then* $\langle \mathfrak{A}(\bar{\sigma}) \rangle \in \mathbb{C}(R)$.

Proof. Since the chain type of $\mathfrak{A}^0 \subseteq \mathfrak{A}^1 \subseteq \cdots$ is eventually in R there exists m such that for all $n \geq m$, $\mathfrak{A}^m \subseteq \cdots \subseteq \mathfrak{A}^n$ has its chain type in R.

Let k be a support for R and let G be the set of all $\mathfrak{A}(\tau)|k$ such that

(a) τ is an index $(\mathfrak{B}^0, b^0)$, ..., $(\mathfrak{B}^s, b^s)$,

(b) $\bar{\sigma}|(m+1) \leq \tau$, and

(c) the chain type of $\mathfrak{B}^r \subseteq \cdots \subseteq \mathfrak{B}^s$, or any such chain obtained by repeating some $\mathfrak{B}^j$, is in R where $\mathfrak{B}^r = \mathfrak{A}^m$.

We claim G is a recursive $^\omega(R|k)$-frame from which $\mathfrak{A}(\bar{\sigma})$ is attainable. Firstly, $G \subseteq \mathrm{MF}|k$. If $\mathfrak{A}(\tau^0)|k \in G$ and $\mathfrak{A}(\tau^1)|k \in G$ then

$$\mathfrak{A}(\tau^0)|k \cap \mathfrak{A}(\tau^1)|k = \mathfrak{A}(\tau^0 \wedge \tau^1)|k.$$

Since τ^0, τ^1 satisfy (b), (c) above so does their common initial segment $\tau^0 \wedge \tau^1$. Hence G is closed under intersections. Clearly $\ulcorner G \urcorner$ is recursively enumerable so $\ulcorner G^* \urcorner$ is and, further, $C_G(B)$ can be computed just as $C_{\mathrm{MF}}(B)$ was computed in the proof of Theorem 38.3.

Suppose $\mathfrak{C}^0 \subset \cdots \subset \mathfrak{C}^s$ is a strictly increasing chain of elements of G then $\mathfrak{C}^i = \mathfrak{A}(\tau|j^i)|k$ for some index τ satisfying (a), (b), (c) above and

$j^i > j^l$ for $i > l$. But then $\tau | j^s$ is an index $(\mathfrak{B}^0, b^0), \dots, (\mathfrak{B}^{j^s}, b^{j^s})$ satisfying (c) so the chain type of $\mathfrak{C}^0 \subset \cdots \subset \mathfrak{C}^s$ is a subchain type of $\mathfrak{A}^0 | k \subset \cdots \subset \mathfrak{A}^{j^s} | k$, indeed of $\mathfrak{A}^m | k \subset \cdots \subset \mathfrak{A}^{j^s} | k$ (by (b)), and hence is in $R | k$. Similarly if some $\mathfrak{C}^j$ is repeated.

Finally, we show $\mathfrak{A}(\bar{\sigma}) | k \in \mathscr{A}(G)$. Since every initial segment of $\bar{\sigma}$ of length $\geq m+1$ is an index τ satisfying (a), (b) and (c), $\{\mathfrak{A}(\bar{\sigma} | r) | k: r = m+1, m+2, \dots\}$ is a chain of elements of G whose union is $\mathfrak{A}(\bar{\sigma}) | k$, therefore $\mathfrak{A}(\bar{\sigma}) | k \in \mathscr{A}(G)$ as claimed. $\square$

Definition 39.3. Let $\mathfrak{A}$ be an infinite strictly increasing chain

$$\mathfrak{A}^0 \subset \mathfrak{A}^1 \subset \cdots$$

of finite objects of $\mathfrak{C}$. Let

$$\sigma = (\tilde{\mathfrak{A}}^0, b^0), \dots, (\tilde{\mathfrak{A}}^n, b^n)$$

be an index then σ is said to be an $\mathfrak{A}$-*index* if for each i, $\tilde{\mathfrak{A}}^i = \mathfrak{A}^i$. Let $F \subseteq \mathfrak{C} | k$ be a frame. Then an index ρ is said to be an F-*stop* if ρ is an $\mathfrak{A}$-index and for no infinite $\mathfrak{A}$-index $\bar{\sigma} \geq \rho$ do we have $\mathfrak{A}(\bar{\sigma}) | k$ attainable from F.

A relation satisfying Definition 37.2 except for condition (1) is said to be *finitary*.

Lemma 39.4. *Let S be a finitary relation on chain types, let k be a support for S and let F be an $^\omega(S | k)$-frame in $\mathfrak{C} | k$. If $\mathfrak{A}^0 \subset \mathfrak{A}^1 \subset \cdots$ is frequently out of S and σ is an $\mathfrak{A}$-index, then there are infinitely many incomparable F-stops $\rho \geq \sigma$.*

Proof. First we observe that if ρ is an F-stop and $\tau \geq \rho$ then τ is an F-stop since if $\bar{\sigma} \geq \tau$ then $\bar{\sigma} \geq \rho$. Second if ρ is an index of length $r+1$ then the infinitely many indices $\rho, (\mathfrak{A}^{r+1}, 0)$; $\rho, (\mathfrak{A}^{r+1}, 1), (\mathfrak{A}^{r+2}, 0)$; $\rho, (\mathfrak{A}^{r+1}, 1), (\mathfrak{A}^{r+2}, 1), (\mathfrak{A}^{r+3}, 0)$; $\dots$ are all $\geq \rho$ and are mutually incomparable. Hence to prove the lemma it suffices to prove the existence of one F-stop $\rho \geq \sigma$.

Suppose $\mathrm{lh}(\sigma) = l$. Since the chain type of $\mathfrak{A}^0 \subset \mathfrak{A}^1 \subset \cdots$ is frequently out of S there exists $m \geq l$ such that the chain type K of $\mathfrak{A}^l \subset \cdots \subset \mathfrak{A}^m$ is not in S. Now if τ is any $\mathfrak{A}$-index $\geq \sigma$ with $\mathrm{lh}(\tau) = m+1$ then

(1) $\mathfrak{A}(\tau | l+1) \subset \cdots \subset \mathfrak{A}(\tau | m+1)$

has chain type K. Now F is an $^\omega(S | k)$-frame so for some r, $l+1 \leq r \leq m+1$, $\mathfrak{A}(\tau | r) | k$ is not in F. Let $\mathfrak{A}(\tau | r) | k$ be the object not in F for which r is a minimum and set

$$\sigma' = \tau | r.$$

Now we find an $\mathfrak{A}$-index $\rho \geq \sigma'$ of length $r+1$ such that
(2) for every $\mathfrak{A}$-index $\rho' \geq \rho$

$$C_F(\mathfrak{A}(\rho) | k) \not\subseteq \mathfrak{A}(\rho') | k.$$

We show that taking ρ either as the index σ', $(\mathfrak{A}', 0)$ or as the index σ', $(\mathfrak{A}', 1)$ will make (2) hold. For suppose not. Then there exist $\mathfrak{A}$-indices $\rho^0 \geq \sigma'$, $(\mathfrak{A}', 0)$ and $\rho^1 \geq \sigma'$, $(\mathfrak{A}', 1)$ such that

$$C_F(\mathfrak{A}(\sigma', (\mathfrak{A}', 0))|k) \subseteq \mathfrak{A}(\rho^0)|k$$

and

$$C_F(\mathfrak{A}(\sigma', (\mathfrak{A}', 1))|k) \subseteq \mathfrak{A}(\rho^1)|k.$$

But then $\sigma \leq \sigma'$, $(\mathfrak{A}', 0) \leq \rho^0$ and $\sigma \leq \sigma'$, $(\mathfrak{A}', 1) \leq \rho^1$ so

$$\begin{aligned} C_F(\mathfrak{A}(\sigma')|k) &\subseteq (\mathfrak{A}(\rho^0)|k) \cap (\mathfrak{A}(\rho^1)|k) \\ &= \mathfrak{A}(\rho^0 \wedge \rho^1)|k \\ &= \mathfrak{A}(\sigma')|k. \end{aligned}$$

This in turn implies $\mathfrak{A}(\sigma')|k \in F$ contrary to the choice of σ'.

We claim ρ is an F-stop. If it is not then there is an infinite $\mathfrak{A}$-index $\bar{\sigma} \geq \rho$ with $\mathfrak{A}(\bar{\sigma})|k$ attainable from F. But $\bar{\sigma} \geq \rho$ so $\mathfrak{A}(\rho) \subseteq \mathfrak{A}(\bar{\sigma})$. But then $\mathfrak{A}(\rho)|k \subseteq \mathfrak{A}(\bar{\sigma})|k$ and $C_F(\mathfrak{A}(\rho)|k) \subseteq \mathfrak{A}(\bar{\sigma})|k$. But $\mathfrak{A}(\rho)|k$ is finite, so $C_F(\mathfrak{A}(\rho)|k)$ is finite and there is therefore an $\mathfrak{A}$-index ρ' such that

$$\rho \leq \rho' \leq \bar{\sigma} \quad \text{and} \quad C_F(\mathfrak{A}(\rho)|k) \subseteq \mathfrak{A}(\rho')|k.$$

This contradicts our choice of ρ (according to (2)) so ρ is indeed an F-stop. $\square$

Let K be a finite chain type then by the *classical type* given by K we mean the (co-ordinatewise) isomorphism type of $\bigcup\{\mathfrak{B}^i: i=0, 1, ...\}$ for any chain $\mathfrak{B}^0 \subseteq \mathfrak{B}^1 \subseteq \cdots$ whose chain type is K.

Clearly this is a good definition for if $\mathfrak{D}^0 \subseteq \mathfrak{D}^1 \subseteq \cdots$ has the same chain type K then for each co-ordinate there exist isomorphisms $p_n^i: \mathfrak{B}_n^i \cong \mathfrak{D}_n^i$ (for all i) which are compatible and such that

$$\bigcup_i p_n^i: \bigcup_i \mathfrak{B}_n^i \cong \bigcup_i \mathfrak{D}_n^i \quad \text{(see Theorem 24.2)}.$$

Now we can state our main theorem. (Here $\mathbb{C}(R)$ is as given by Definition 37.3.)

Theorem 39.5. *Let K be a strictly increasing finite $(\omega+1)$-chain type of $\mathbb{C}$. Then there exists a sequence $Z = \{Z_i: i=0, 1, ...\}$ such that $Z_i \in \Lambda(\mathbb{C}_i)$, Z has the classical type given by K and, for all $R \in \mathscr{L}(\mathbb{C})$,*

$$Z \in \mathbb{C}(R) \quad \text{if, and only if, K is eventually in } R.$$

Proof. Let $\mathfrak{A}^0 \subseteq \mathfrak{A}^1 \subseteq \cdots$ have chain type K.

Let $F^0, F^1, ...$ be a list of all those recursive $^\omega(S|k)$ frames for which k is a support of S, $S \in \mathscr{L}(\mathbb{C})$ and K is frequently out of S. By using Lemma 39.4

infinitely often we can define an infinite sequence of indices $\sigma^0, \sigma^1, \ldots$ such that σ^n is an F^n-stop extending the index σ^{n-1} (let σ^{-1} be arbitrary). Let $\bar{\sigma}$ be the infinite index extending all the σ^n. Let $\mathfrak{Z}_i$ be the i-th co-ordinate of $\mathfrak{A}(\bar{\sigma})$ and let $\mathsf{Z} = \langle \mathfrak{Z}_i \rangle$. By Lemma 39.2 $\langle \mathfrak{A}(\bar{\sigma}) \rangle \in \mathbb{C}(R)$ for all R such that K is eventually in R provided $\mathfrak{A}(\bar{\sigma})$ is Dedekind. By Definition 39.3 if K is frequently out of S $\langle \mathfrak{A}(\bar{\sigma}) \rangle \notin \mathbb{C}(S)$. So to complete the proof we must show $\mathfrak{A}(\bar{\sigma})$ is Dedekind.

(*Note.* We could have built into our construction of $\bar{\sigma}$ an additional clause to make $\mathfrak{A}(\bar{\sigma})$ Dedekind but this is not necessary as $\mathfrak{A}(\bar{\sigma})$ has already been made sufficiently nondescript.)

Suppose $\mathfrak{A}(\bar{\sigma})$ is not Dedekind then let i be the smallest co-ordinate such that $|\mathfrak{A}(\bar{\sigma})_i|$ contains an infinite recursively enumerable subset B and let $A = |\mathfrak{A}(\bar{\sigma})_i|$. By Theorem 35.7 since the master frame is a tree frame A is regressive. Since B as a set of natural numbers is cofinal in A we can recursively enumerate A by using a regressing function for A on B. Hence we can find an infinite strictly increasing chain of finite objects in $\mathbb{C}_i$

$$\mathfrak{B}^0 \subset \mathfrak{B}^1 \subset \cdots$$

in a completely effective way such that $\bigcup \mathfrak{B}^j = \mathfrak{A}(\bar{\sigma})_i$. Let R^0 be the recursively enumerable finitary relation of support $i+1$ such that a chain type L is in $R^0 | i+1$ if, and only if, L_i is the chain type of a finite subchain of the inclusion chain

$$\mathfrak{B}^0 \subset \mathfrak{B}^2 \subset \mathfrak{B}^4 \subset \cdots.$$

Let $G = {}^\circ\mathbb{C}_0 \times \cdots \times {}^\circ\mathbb{C}_{i-1} \times \{\mathfrak{B}^0, \mathfrak{B}^2, \mathfrak{B}^4, \ldots\}$. Then clearly G is a recursive $R^0 | i+1$ frame from which $\mathfrak{A}(\bar{\sigma}) | (i+1)$ is attainable.

The given chain type K is eventually in R^0. For if it were not, then G would be one of the F^n and there would be an F^n-stop (that is, a G-stop) $\sigma \le \bar{\sigma}$ so $\mathfrak{A}(\bar{\sigma}) | i+1$ would not be attainable from G.

In exactly the same way let R^1 be the finitary relation of support $i+1$ such that a chain type L is in R^1 if, and only if, L_i is the chain type of a finite subchain of the inclusion chain

$$\mathfrak{B}^1 \subset \mathfrak{B}^3 \subset \mathfrak{B}^5 \subset \cdots.$$

Then K is eventually in both R^0 and R^1. But R^0 and R^1 are closed under the formation of subchain types so some non-empty subchain of K is in $R^0 \cap R^1$. But this intersection is empty since $\mathfrak{B}^0 \subset \mathfrak{B}^1 \subset \cdots$ is strictly increasing; so we have a contradiction and $\mathfrak{A}(\bar{\sigma})_i$ is Dedekind for all i. This completes the proof. $\square$

Corollary 39.6. *With the same hypotheses as the theorem there are c sequences Z of regressive Dedekind types satisfying the conclusion of the theorem.*

Proof. Since the master frame MF is a tree frame all $(Z_0, ..., Z_{k-1})$ attainable from MF$|k$ are regressive k-tuples by Theorem 35.7. We sketch two proofs of the rest of the corollary.

1. Since any $\mathfrak{A}$-index σ has two incomparable F^n-stops above it, we have a binary tree in the master frame MF, all of whose infinite branches give regressive k-tuples of the appropriate type.

2. Topologize infinite $\mathfrak{A}$-indices by taking as basic open sets those of the form $\{\bar{\sigma}: \bar{\sigma} \geq \sigma\}$ for σ a finite index. Then the $\mathfrak{A}(\bar{\sigma})$ which we seek are those with $\bar{\sigma}$ in a co-meagre set; but any such set has cardinality c. This completes the proof. $\square$

We remark that at the cost of more detail we can also require that for each i with Z_i infinite, all the continuum many Z differ at the i-th coordinate.

40. Compactness and Dimension

In this final section we consider a weak infinite product category $\mathbb{C}$ where each $\mathbb{C}_i$ has dimension. As in Section 26 consideration of chain types reduces to consideration of 0-chain types only and then, since we have dimension, to natural numbers. Thus our concern here is the extension of relations on natural numbers to relations on Dedekind types.

Let $\times_\omega E$ be the set of all finitely non-zero sequences of natural numbers and $\times_\omega E^+$ be the set of all similar sequences with elements in $E \cup \{\aleph_0\}$. Then let dim be the function induced by the dimension functions of the $\mathbb{C}_i$ thus

$$\dim: \mathbb{C} \to \times_\omega E^+ \quad \text{and} \quad \dim \mathfrak{A} = (\dim_0 \mathfrak{A}_0, \dim_1 \mathfrak{A}_1, ...)$$

where $\dim_i: \mathbb{C}_i \to E^+$.

Let $\mathbf{R} \subseteq \times_\omega E$ then $\mathbf{R}$ is a (recursively enumerable) finitary relation if there exists a number k such that ($\mathbf{R}$ is recursively enumerable and) for all $x \in \times_\omega E$, $x \in \mathbf{R}$ if and only if $x|k \in \mathbf{R}|k$.

A finitary relation $\mathbf{R} \subseteq \times_\omega E$ naturally corresponds to a finitary relation R on finite types ($=$0-chain types) under the correspondence that the 0-chain type of $\mathfrak{A}$ is in R if and only if dim $\mathfrak{A}$ is in $\mathbf{R}$. Thus with each recursively enumerable finitary relation $\mathbf{R} \subseteq \times_\omega E$ of support k we may associate

$$\mathbb{C}(\mathbf{R}) = \{X \in \Lambda(\mathbb{C}): X|k \in \mathbb{C}(^\circ R|k)\}$$

where $R = \{\langle \mathfrak{A} \rangle: \dim \mathfrak{A} \in \mathbf{R}\}$. So $X \in \mathbb{C}(\mathbf{R})$ if and only if $X|k$ is attainable from a recursive frame in $\mathbb{C}|k$ whose members all have their dimensions in $\mathbf{R}|k$.

Let $\mathscr{L}(E)$ be the lattice of recursively enumerable finitary relations in $\times_\omega E$. A *filter* $\mathscr{F} \subseteq \mathscr{L}(E)$ is a non-empty subset of $\mathscr{L}(E)$ such that

(i) $\mathbf{R}, \mathbf{S} \in \mathcal{F}$ implies $\mathbf{R} \cap \mathbf{S} \in \mathcal{F}$, (ii) $\mathbf{R} \in \mathcal{F}$ and $\mathbf{S} \in \mathcal{L}(E)$ and $\mathbf{R} \subseteq \mathbf{S}$ imply $\mathbf{S} \in \mathcal{F}$. A filter $\mathcal{F}$ is *proper* if $\emptyset \notin \mathcal{F}$ and is a *prime* filter if $\mathbf{R} \cup \mathbf{S} \in \mathcal{F}$ implies $\mathbf{R} \in \mathcal{F}$ or $\mathbf{S} \in \mathcal{F}$.

Let $\mathbf{C}$ be the set of those natural numbers j such that $\{y \in \times_\omega E : y_j = n\}$ is in $\mathcal{F}$ for some n. Let a_j be the unique such n corresponding to j for $j \in \mathbf{C}$. (n is unique since $\mathcal{F}$ is proper.)

In Section 39 we needed a chain to allow us to characterize sets whose extension contained a fixed sequence of Dedekind types. In our present situation the role of the chain may be replaced by a prime filter as we show below.

Recall that if $x, y \in \times_\omega E$ then $x < y$ means $x_i \leq y_i$ for all i but $x \neq y$.

Theorem 40.1. *Let $\mathcal{F}$ be a proper prime filter of $\mathcal{L}(E)$ such that $\mathbf{C} \neq E$. Then there is a strictly increasing sequence $x^0 < x^1 < \cdots$ of elements of $\times_\omega E$ such that for all $\mathbf{R} \in \mathcal{F}$ there is a natural number n such that for all $m \geq n$, $x^m \in \mathbf{R}$.*

We need a Lemma.

Lemma 40.2. *Suppose $\mathbf{R} \in \mathcal{F}$. Let k be a support for $\mathbf{R}$ and let b be a natural number. Then there exists $z \in \mathbf{R}$ such that $z_j = a_j$ for $j \in k \cap \mathbf{C}$ and $z_j \geq b$ for $j \in k - \mathbf{C}$.*

Proof. First we observe that

$$\{y \in \times_\omega E : y_j \geq b\} \cup \bigcup_{0 \leq i < b} \{y \in \times_\omega E : y_j = i\} = \times_\omega E \in \mathcal{F}.$$

If $j \in k - \mathbf{C}$ then $\{y \in \times_\omega E : y_j = i\} \notin \mathcal{F}$ for any i by the definition of $\mathbf{C}$. Since $\mathcal{F}$ is prime, $\{y \in \times_\omega E : y_j \geq b\} \in \mathcal{F}$. But then each of the finitely many sets $\mathbf{R}$, $\{y \in \times_\omega E : y_j = a_j\}$, for $j \in k \cap \mathbf{C}$, and $\{y \in \times_\omega E : y_j \geq b\}$, for $j \in k - \mathbf{C}$, is in $\mathcal{F}$ so their intersection is in $\mathcal{F}$. In particular the intersection is non-empty. Any z in this intersection has the properties required for the lemma. $\square$

Now we can complete the proof of the theorem. Let $\mathbf{R}^0, \mathbf{R}^1, \ldots$ be an enumeration of $\mathcal{F}$ and choose k^i to be a support for $\mathbf{R}^i$ so that $k^0 < k^1 < \cdots$ and $k^0 >$ some $c \in E - \mathbf{C}$. By Lemma 40.2 there exists $x^0 \in \mathbf{R}^0$ such that $x_j^0 = a_j$ for $j \in k^0 \cap \mathbf{C}$ and $x_j^0 = 0$ for $j \geq k^0$. Now we can recursively choose x^{n+1} by the lemma. Take $\mathbf{R}$ as $\mathbf{R}^0 \cap \cdots \cap \mathbf{R}^{n+1}$ then k^{n+1} is a support for $\mathbf{R}$ since the k^m are strictly increasing. Take

$$b > \max \{x_j^m : m \leq n, j \in k^n - \mathbf{C}\}.$$

Let z be given by the lemma and let $x_j^{n+1} = z_j$ for $j < k^{n+1}$, $x_j^{n+1} = 0$ otherwise.

Clearly $x^{n+1} > x^n$ for each n. Since every element of $\mathcal{F}$ is some $\mathbf{R}^n$ and $x^n \in \mathbf{R}^n$ the theorem follows. $\square$

Let $\mathcal{R}$ be a set of recursively enumerable finitary relations $\mathbf{R} \subseteq \times_\omega E$. We say that $x \in \times_\omega E$ satisfies $\mathcal{R}$ if $x \in \mathbf{R}$ for every $\mathbf{R} \in \mathcal{R}$. Similarly if $\mathbb{C}(\mathcal{R}) = \{\mathbb{C}(\mathbf{R}): \mathbf{R} \in \mathcal{R}\}$ is the set of extensions to Dedekind types of such a set of recursively enumerable finitary relations we say that X, where $X_i \in \Lambda(\mathbb{C}_i)$, satisfies the set of extensions to Dedekind types of $\mathcal{R}$ if $X \in \mathbb{C}(\mathbf{R})$ for every $\mathbf{R} \in \mathcal{R}$.

Theorem 40.3 (The Compactness Theorem). *Let $\mathcal{R}$ be a set of recursively enumerable finitary relations $\mathbf{R} \subseteq \times_\omega E$.*

If every finite subset of $\mathcal{R}$ is satisfied by some element of $\times_\omega E$ then the set of extensions to Dedekind types of $\mathcal{R}$ is satisfied by some Z with $Z_i \in \Lambda(\mathbb{C}_i)$.

Proof. Extend $\mathcal{R}$ to a prime filter $\mathcal{F}$ and let $\mathbf{C}$ be chosen as for Theorem 40.1. If $\mathbf{C} = E$, the theorem is trivial. If this is not the case, then we let $\mathbf{R}^0, \mathbf{R}^1, \ldots$ be an enumeration of $\mathcal{F}$. Choose $x^0 < x^1 < \cdots$ by Theorem 40.1 and let $\mathfrak{A}^0 \subset \mathfrak{A}^1 \subset \cdots$ be a corresponding strictly increasing chain of finite objects of $\mathbb{C}$ such that $\dim_i(\mathfrak{A}_i^j) = x_i^j$. By Theorem 39.5 there exists Z such that for all recursively enumerable finitary relations R on chain types if $\mathfrak{A}^0 \subset \mathfrak{A}^1 \subset \cdots$ is eventually in R then $Z \in \mathbb{C}(R)$. We claim that Z has the properties by the theorem.

For each $\mathbf{R} \in \mathcal{R}$ let R be the set of all chain types of inclusion chains whose members have dimensions in $\mathbf{R}$. If $\mathbf{R}$ is recursively enumerable finitary, so is R. Moreover, $\mathbf{R}^i \in \mathcal{F}$ implies that the type of $\mathfrak{A}^0 \subset \mathfrak{A}^1 \subset \cdots$ is eventually in R^i due to the choice of $x^0 < x^1 < \cdots$ by Theorem 40.1. So $Z \in \mathbb{C}(R^i)$ due to the choice of Z by Theorem 39.5. This concludes the proof. $\square$

Now we can derive a powerful corollary about equations. For notational simplicity we treat only the case when all the $\mathbb{C}_i$ are the same category $\mathbb{C}'$. Then $\mathbb{C}$ is the weak infinite power of $\mathbb{C}'$. We also assume $\mathbb{C}'$ has dimension and degree 1. We use and interpret formal terms $t^1, t^2, \ldots, s^1, s^2, \ldots$ involving only variables from an infinite list $v_0, v_1, \ldots$ as in Section 28. Since each term involves only a finite number of variables, if $s = t$ is an equation with variables among $v_0, \ldots, v_n$ then we interpret s, t as finitary recursive strict combinatorial functors from $\mathbb{C}' \times \cdots \times \mathbb{C}' = \mathbb{C} \mid n$ into $\mathbb{C}'$ and, for natural numbers, as the strict combinatorial functions induced from $\times^n E$ to E. Similarly for the other interpretations.

Corollary 40.4. *If*

$$(1) \qquad (t^0 = t^1 \mathrel{\&} t^2 = t^3 \mathrel{\&} \ldots) \mathrel{\&} (s^0 \neq s^1 \mathrel{\&} s^2 \neq s^3 \mathrel{\&} \ldots)$$

is satisfiable in the natural numbers then it is satisfiable in $\mathbb{C}'$-Dedekind types.

Proof. Let $\mathbf{T}^i = \{x \in \times_\omega E : x$ satisfies $t^{2i} = t^{2i+1}\}$ and $\bar{\mathbf{S}}^i = \{x \in \times_\omega E :$ x satisfies $s^{2i} \neq s^{2i+1}\}$. Set $\mathbf{S}^i = \times_\omega E - \bar{\mathbf{S}}^i$. By Theorem 40.3 there is an infinite sequence Z of Dedekind $\mathbb{C}'$-types such that for all i, $Z \in \mathbb{C}(\mathbf{T}^i)$ and $Z \in \mathbb{C}(\bar{\mathbf{S}}^i)$. Now $\mathbf{S}^i \cap \bar{\mathbf{S}}^i = \emptyset$ implies $\mathbb{C}(\mathbf{S}^i) \cap \mathbb{C}(\bar{\mathbf{S}}^i) = \emptyset$ follows from Corollary 27.2 and the fact that $\mathbb{C}(\emptyset) = \emptyset$ (since the only $\emptyset$-frame is the empty set). Hence, for all i, $Z \notin \mathbb{C}(\mathbf{S}^i)$.

Now suppose t^{2i}, t^{2i+1} have all their variables among $v_0, \ldots, v_{n-1}$ then Z satisfies $t^{2i} = t^{2i+1}$ if, and only if, $Z|n$ satisfies $t^{2i} = t^{2i+1}$ (in $\Lambda(\mathbb{C}|n)$) if, and only if, $Z|n \in (\mathbb{C}|n)(\mathbf{T}^i|n)$ by Theorem 26.1 if, and only if, $Z \in \mathbb{C}(\mathbf{T}^i)$ by definition of $\mathbb{C}(\mathbf{T}^i)$. Similarly Z satisfies $s^{2i} = s^{2i+1}$ if, and only if, $Z \in \mathbb{C}(\mathbf{S}^i)$. So Z satisfies the conjunction (1) as required. $\quad\square$

Now we can prove Nerode's result [1966] that countable diophantine correct models of formal arithmetic are embeddable in the $\mathbb{S}$-Dedekind types.

Definition 40.5. A model $\mathfrak{A} = (A, +, \cdot)$ of formal arithmetic is said to be *diophantine correct* if every diophantine equation which has a solution in $\mathfrak{A}$ has a solution in the natural numbers.

By the compactness theorem for first order logic there are many models elementarily equivalent to the (full structure of the) natural numbers. All of these are diophantine correct since we can formalize the statement that a given equation has a solution. On the other hand, Matijasevich [1970] via Julia Robinson has shown that Hilbert's 10th problem is recursively unsolvable. So the set of diophantine equations without solutions is not recursively enumerable and in particular the recursively enumerable set of axioms of formal arithmetic cannot yield an effective enumeration of those without solutions. Hence it is consistent to add to the axioms of formal arithmetic that certain diophantine equations which do not have solutions in the natural numbers do have solutions. That is, there exist diophantine incorrect models of formal arithmetic.

Corollary 40.6 (Theorem 5.2 of Nerode [1966]). *Any countable diophantine correct model $\mathfrak{A} = (A, +, \cdot)$ of formal arithmetic is embeddable in $(\Lambda(\mathbb{S}), +, \cdot)$.*

Proof. We may assume E is an initial segment of $\mathfrak{A}$. Let $a_0, a_1, \ldots$ be a list of all elements of A. Let $D(\mathfrak{A})$ be the set consisting of the following formulae

$$
\begin{array}{lll}
v_i + v_j = v_k & \text{for all } i, j, k \text{ such that} & a_i + a_j = a_k \\
v_i \cdot v_j = v_k & \text{for all } i, j, k \text{ such that} & a_i \cdot a_j = a_k \\
v_i + v_j \neq v_k & \text{for all } i, j, k \text{ such that} & a_i + a_j \neq a_k \\
v_i \cdot v_j \neq v_k & \text{for all } i, j, k \text{ such that} & a_i \cdot a_j \neq a_k \\
v_i \neq v_j & \text{for all } i, j \quad\; \text{such that} & a_i \neq a_j .
\end{array}
$$

We now show that every finite subset of $D(\mathfrak{A})$ is satisfiable in E. The conjunction of such a finite subset is of the form

$$(2) \qquad p^0 = p^1 \& \ldots \& p^{2i} = p^{2i+1} \& q^0 \neq q^1 \& \ldots \& q^{2j} \neq q^{2j+1}.$$

Now if p, p', q, q' are polynomials with coefficients from E then $p = p' \& q = q'$ may be replaced by $p^2 + (p')^2 + q^2 + (q')^2 = 2pp' + 2qq'$ (where superscripts denote powers) and an inequality $q \neq q'$ between polynomials may be replaced by $\exists v_k (q^2 + (q')^2 = 1 + 2qq' + v_k)$ (where superscripts denote powers) provided v_k does not occur in q, q'. Hence (2) is equivalent in formal arithmetic to a formula

$$(3) \qquad\qquad\qquad (\exists \mathbf{v})\,(p = p')$$

where p, p' are polynomials and $\mathbf{v}$ is a finite string of variables. Since $\mathfrak{A}$ is diophantine correct, if (3) is satisfiable in $\mathfrak{A}$ it is satisfiable in E. Hence every finite subset of $D(\mathfrak{A})$ is satisfiable in E. By the Compactness Theorem 40.3, $D(\mathfrak{A})$ is satisfiable in $\Lambda(\mathbb{S})$ by Z, say. But then the map $a_i \rightsquigarrow Z_i$ is the required embedding. $\square$

We conclude with a stronger version of the Compactness Theorem.

Definition 40.7. If $R, S \in \mathscr{L}(E)$ then R is said to be *eventually contained in* S (written $R \subseteq_e S$) if there is a natural number n such that if k is a support for both R and S, $x \in R$ and $x_0 > n, \ldots, x_{k-1} > n$ then $x \in S$.

Note that $R \subseteq S$ trivially implies $R \subseteq_e S$.

Definition 40.8. A filter $\mathscr{F}$ is said to be a *realizability filter* if (i) $\mathscr{F}$ is proper and (ii) $R \in \mathscr{F}$, $S \in \mathscr{L}(E)$ and $R \subseteq_e S$ imply $S \in \mathscr{F}$.

The next theorem shows how realizability filters arise.

Theorem 40.9. Let $X \in \times^\omega \Lambda^\infty(\mathbb{C}_i)$. Let $\mathscr{F}$ consist of all $R \in \mathscr{L}(E)$ such that $X \in \mathbb{C}(R)$. Then $\mathscr{F}$ is a realizability filter.

Proof. Trivially $X \in \mathbb{C}(E)$ so $\mathscr{F}$ is non-empty. If $R, S \in \mathscr{F}$ then $R \cap S \in \mathscr{F}$ since $\mathbb{C}(R \cap S) = \mathbb{C}(R) \cap \mathbb{C}(S)$ follows from Corollary 27.2(i). (That $\mathscr{F}$ is a filter follows from Corollary 27.2(ii) but we need a stronger result.)

If $R \in \mathscr{F}$, $R \subseteq_e S$, $X \in \mathbb{C}(R)$ and k is a support for both R and S then there exists $\mathfrak{X} = (\mathfrak{X}_0, \ldots, \mathfrak{X}_{k-1})$ such that $X_i = \langle \mathfrak{X}_i \rangle$ for $i < k$ and a recursive $R|k$-frame G such that $\mathfrak{X}$ is attainable from G. Since $R \subseteq_e S$ let n be a number given according to Definition 40.7 and let $\mathfrak{Y}$ be any element of G such that $\mathfrak{Y} \subseteq \mathfrak{X}$ and $\dim \mathfrak{Y}_i > n$ for $i < k$. Let $H = \{\mathfrak{Z} \in G : \mathfrak{Y} \subseteq \mathfrak{Z}\}$ then clearly H is a recursive frame from which $\mathfrak{X}$ is attainable. Moreover, H is an S-frame by the choice of the $\mathfrak{Y}_i$. Hence $X \in \mathbb{C}(S)$ so $S \in \mathscr{F}$. $\square$

In fact all realizability filters arise in this way.

Theorem 40.10. Let $\mathscr{F}$ be a realizability filter in $\mathscr{L}(E)$. Then there exists $Z \in \times^\omega \Lambda^\infty(\mathbb{C}^i)$ such that for all $R \in \mathscr{L}(E)$,

$$Z \in \mathbb{C}(R) \quad \textit{if, and only if,} \quad R \in \mathscr{F}.$$

Proof. Essentially all we have to do is throw away all the finite Z_i from Theorem 40.3. We need

Theorem 40.11. *Let $\mathscr{F}$ be a realizability filter in $\mathscr{L}(E)$. Then there is a strictly increasing sequence $x^0 < x^1 < \cdots$ of elements of $\times_\omega E$ such that for all $R \in \mathscr{L}(E)$, $R \in \mathscr{F}$ if, and only if, there is a natural number n such that for all $m \geq n$, $x^m \in R$. Moreover, for each i, $\{x_i^m: m = 0, 1, 2, \ldots\}$ is unbounded.*

Proof. We proceed as for Theorem 40.1. Since $\mathscr{F}$ is a realizability filter and is proper $R \nsubseteq_e \emptyset$ for any $R \in \mathscr{F}$. Since $\{y \in \times_\omega E: y_j = n\} \subseteq_e \emptyset$ we have that for every n and j, $\{y \in \times_\omega E: y_j = n\} \notin \mathscr{F}$. We now need a lemma corresponding to Lemma 40.2 but this time the proof is easier.

Lemma 40.12. *Suppose $R \in \mathscr{F}$ and $S \in \mathscr{L}(E) - \mathscr{F}$. Let k be a support for R and S and let b be a natural number. Then there exists $z \in R - S$ such that $z_j \geq b$ for all $j < k$.*

Proof. Immediate since $R \nsubseteq_e S$. □

To complete the proof of Theorem 40.11 we let $S^0, S^1, \ldots$ be an enumeration of $\mathscr{L}(E) - \mathscr{F}$ such that each $S \in \mathscr{L}(E) - \mathscr{F}$ is repeated infinitely often, and let $R^0, R^1, \ldots$ be an enumeration of $\mathscr{F}$. Now let k^i be a support for R^i and S^i chosen so that $k^{j+1} > k^j$ for all j. By Lemma 40.12 there exists $x^i \in (R^0 \cap \cdots \cap R^i) - S^i$ such that $x_j^i > \max \{x_l^{i-1}: l < k^{i-1}\}$ for all $j < k^i$ and $x_j^i = 0$ for $j \geq k^i$. Clearly $\{x_j^i: i = 0, 1, 2, \ldots\}$ is unbounded. □

Now we can complete the proof of Theorem 40.10 exactly as we proved the Compactness Theorem 40.3 (using Theorem 39.5). The only extra thing we need is that each Z_i is infinite but this is immediate as $\{x_j^i: i = 0, 1, 2, \ldots\}$ is unbounded for each j. This completes the proof. □

Finally we observe that Theorem 40.10 yields the Dedekind type required for Theorems 29.1 and 29.2 on universal sentences; and in fact can be used to strengthen those theorems.

Corollary 40.13 (In the dimension case). *Let X be such that $X_i \in \Lambda(\mathbb{C}_i)$. Then there exists Y with every $Y_i \in \Lambda(\mathbb{C}_i)$ such that for all n, $Y \mid n$ is regressive and for all $R \in \mathscr{L}(E)$, $Y \in \mathbb{C}(R)$ if, and only if, $X \in \mathbb{C}(R)$.*

Proof. Let $Y_i = X_i$ for those i such that X_i is finite. Then consider the weak infinite product omitting those co-ordinates i. Then every X_i is infinite. By Theorem 40.9 there is a realizability filter $\mathscr{F}$ such that $R \in \mathscr{F}$ if, and only if, $X \in \mathbb{C}(R)$. By Theorem 40.10 there exists Y such that $Y \in \mathbb{C}(R)$ if, and only if, $R \in \mathscr{F}$. But $Y \mid n$ may be taken to be regressive by Corollary 39.6. □

Bibliography

Aczel, P. H. G.: [1966] D. Phil. thesis Oxford (1966).

Baldwin, J. T., Lachlan, A. H.: [1971] On strongly minimal sets. J. Symbolic Logic **36**, 79–96 (1971).

Bell, J. L., Slomson, A. B.: [1969] Models and Ultraproducts. Amsterdam (1969).

Burnside, W.: [1883] Theory of Groups of Finite Order. Cambridge (1883).

Crossley, J. N.: [1963] D. Phil. Thesis, Oxford (1963).

[1965] Constructive order types I, in Formal Systems and Recursive Functions. Amsterdam (1965).

[1969] Constructive order types. Amsterdam (1969).

[1970] Recursive Equivalence. Bull. L. M. S. **2**, 129–151 (1970).

Dekker, J. C. E.: [1955] A non-constructive extension of the number system. J. Symbolic Logic **20**, 204–205 (1955).

[1966] Les Fonctions combinatoires et les isols. Paris (1966).

[1969] Countable vector spaces with recursive operations, Part I. J. Symbolic Logic **34**, 363–387 (1969).

Dekker, J. C. E., Myhill, J.: [1960] Recursive equivalence types. Univ. of California publications in mathematics, n. s. **3**, 67–214 (1960).

Ellentuck, E.: [1965] The universal properties of Dedekind finite cardinals. Ann. of Math. **82**, 225–248 (1965).

[1966] Generalized idempotence in cardinal arithmetic. Fund. Math. **58**, 241–258 (1966).

[1969] A choice free theory of Dedekind cardinals. J. Symbolic Logic **34**, 70–84 (1969).

[1972] The positive properties of isolic integers. J. Symbolic Logic **37**, 114–132 (1972).

Hamilton, A. G.: [1970] Bases and α-dimensions of countable vector spaces with recursive operations. J. Symbolic Logic **35**, 85–96 (1970).

Hassett, M. J.: [1964] Ph. D. thesis Rutgers (1964).

Hay, L. S.: [1966] The co-simple isols. Ann. of Math. **83**, 231–256 (1966).

Jónsson, B.: [1962] Algebraic extensions of relational systems. Math. Scand. **11**, 179–205 (1962).

Kelley, J. L.: [1955] General Topology. Princeton (1955).

Marsh, W.: [1966] Ph. D. disertation. Dartmouth (1966).

Matijasevich, Yu. V.: [1970] Diofantovost perechislimekh mioshestv. Dokl. Akad. Nauk SSSR, **191**, no. 2, 279–282 (1970); English translation: Soviet Math. Dokl. **11**, No. 2, 354–357 (1970).

Mendelson, E.: [1964] Introduction to Mathematical Logic. Princeton (1964).

Morley, M. D.: [1965] Categoricity in power. Trans. Amer. Math. Soc. **114**, 514–538 (1965).

[1967] Countable models of $\aleph_1$-categorical theories. Israel J. Math. **5**, 65–72 (1967).

Myhill, J.: [1958] Recursive equivalence types and combinatorial functions. Bull. Amer. Math. Soc. **64**, 373–376 (1958).

Nerode, A.: [1961] Extensions to isols. Ann. of Math. **73**, 362–403 (1961).
  [1966] Diophantine correct non-standard models in the isols. Ann. of Math. **84**, 421–432 (1966).
  [1966a] Combinatorial series and recursive equivalence types. Fund. Math. **58**, 133–141 (1966).
Sacks, G. E.: [1972] Saturated Model Theory. New York (1972).
Shelah, S.: [1971] Stability, the f.c.p. and superstability. Annals of Math. Logic **3**, 271–362 (1971).
Vaught, R.L.: [1959] Denumerable models of complete theories in Infinitistic Methods. Warsaw (1961).
van der Waerden, B.L.: [1931] Moderne Algebra. Berlin (1931).

An extensive bibliography of Recursive Equivalence will be found in Crossley [1970].
Addenda to that bibliography will be found in the Recursive Function Theory Newsletter 1974.

Index of Notations

We here list the notations used in order of appearance. Standard notations have been used as far as possible but a brief explanation is included for clarification.

Note that subscripts denote co-ordinates, superscripts are used for indexing.

$\subset$ always denotes *strict* inclusion so $A \subset B$ if, and only if, $A \subseteq B \& A \neq B$.

E denotes the set of natural numbers $\{0, 1, 2, ...\}$ and $\emptyset$ the empty set.

In the list below a brief explanation of definitions is given for each notation. Standard logical notation $\&$, $\vee$, $\neg$, $\rightarrow$, $\exists$, $\forall$ is used.

General Index

Ergebnisse der Mathematik und ihrer Grenzgebiete